INVESTIGATIONS IN NUMBER, DATA, AND SPACE®

Geometry and Fractions

Shapes, Halves, and Symmetry

Grade 2

Also appropriate for Grade 3

Joan Akers
Michael T. Battista
Anne Goodrow
Douglas H. Clements
Julie Sarama

Developed at TERC, Cambridge, Massachusetts

Dale Seymour Publications®
White Plains, New York

The *Investigations* curriculum was developed at TERC (formerly
Technical Education Research Centers) in collaboration with Kent State
University and the State University of New York at Buffalo. The work was
supported in part by National Science Foundation Grant No. ESI-9050210.
TERC is a nonprofit company working to improve mathematics and science
education. TERC is located at 2067 Massachusetts Avenue, Cambridge,
MA 02140.

**This project was supported, in part,
by the**
National Science Foundation
Opinions expressed are those of the authors
and not necessarily those of the Foundation

Managing Editor: Catherine Anderson
Grade-Level Editor: Alison Abrohms
Series Editor: Beverly Cory
Revision Team: Laura Marshall Alavosus, Ellen Harding, Patty Green Holubar,
Suzanne Knott, Beverly Hersh Lozoff
ESL Consultant: Nancy Sokol Green
Production/Manufacturing Director: Janet Yearian
Production/Manufacturing Coordinator: Joe Conte
Design Manager: Jeff Kelly
Design: Don Taka
Illustrations: Laurie Harden, Susan Jaekel, Meryl Treatner
Cover: Bay Graphics
Composition: Archetype Book Composition

This book is published by Dale Seymour Publications®, an imprint of
Addison Wesley Longman, Inc.

Dale Seymour Publications
10 Bank Street
White Plains, NY 10602
Customer Service: 1-800-872-1100

**DALE
SEYMOUR
PUBLICATIONS®**

Order number DS43803
ISBN 1-57232-656-5
10 11 12 13 14 15-ML-06 05 04 03

 Printed on Recycled Paper

INVESTIGATIONS IN NUMBER, DATA, AND SPACE®

TERC

Principal Investigator   Susan Jo Russell

Co-Principal Investigator   Cornelia Tierney

Director of Research and Evaluation   Jan Mokros

Director of K–2 Curriculum   Karen Economopoulos

Curriculum Development
Joan Akers
Michael T. Battista
Mary Berle-Carman
Douglas H. Clements
Karen Economopoulos
Anne Goodrow
Marlene Kliman
Jerrie Moffett
Ricardo Nemirovsky
Andee Rubin
Susan Jo Russell
Cornelia Tierney
Tracey Wright

Evaluation and Assessment
Mary Berle-Carman
Jan Mokros
Andee Rubin

Teacher Support
Anne Goodrow
Liana Laughlin
Jerrie Moffett
Megan Murray
Tracey Wright

Technology Development
Michael T. Battista
Douglas H. Clements
Julie Sarama

Video Production
Megan Murray
David A. Smith
Judy Storeygard

Administration and Production
Irene Baker
Amy Catlin
Amy Taber

*Cooperating Classrooms
for This Unit*
Rose Christiansen
*Brookline Public Schools
Brookline, MA*

Lisa Seyferth
Carol Walker
*Newton Public Schools
Newton, MA*

Phyllis Ollove
*Boston Public Schools
Boston, MA*

Margaret McGaffigan
*Nashoba Regional School District
Stow, MA*

Barbara Rynerson
Dale Dhoore
Beth Newkirk
Pat McLure
*Oyster River Public Schools
Durham, NH*

Consultants and Advisors
Deborah Lowenberg Ball
Marilyn Burns
Ann Grady
James J. Kaput
Mary M. Lindquist
John Olive
Leslie P. Steffe
Grayson Wheatley

Graduate Assistants
Kathryn Battista
Caroline Borrow
Judy Norris
Kent State University

Julie Sarama
Sudha Swaminathan
Elaine Vukelic
State University of New York at Buffalo

Revisions and Home Materials
Cathy Miles Grant
Marlene Kliman
Margaret McGaffigan
Megan Murray
Kim O'Neil
Andee Rubin
Susan Jo Russell
Lisa Seyferth
Myriam Steinback
Judy Storeygard
Anna Suarez
Cornelia Tierney
Carol Walker
Tracey Wright

CONTENTS

TEACHER NOTES

WHERE TO START

The first-time user of *Shapes, Halves, and Symmetry* should read
the following:

When you next teach this same unit, you can begin to read more of the
background. Each time you present the unit, you will learn more about
how your students understand the mathematical ideas.

Investigations in Number, Data, and Space® is a K–5 mathematics curriculum with four major goals:

- to offer students meaningful mathematical problems
- to emphasize depth in mathematical thinking rather than superficial exposure to a series of fragmented topics
- to communicate mathematics content and pedagogy to teachers
- to substantially expand the pool of mathematically literate students

The *Investigations* curriculum embodies a new approach based on years of research about how children learn mathematics. Each grade level consists of a set of separate units, each offering 2–8 weeks of work. These units of study are presented through investigations that involve students in the exploration of major mathematical ideas.

Approaching the mathematics content through investigations helps students develop flexibility and confidence in approaching problems, fluency in using mathematical skills and tools to solve problems, and proficiency in evaluating their solutions. Students also build a repertoire of ways to communicate about their mathematical thinking, while their enjoyment and appreciation of mathematics grows.

The investigations are carefully designed to invite all students into mathematics—girls and boys, members of diverse cultural, ethnic, and language groups, and students with different strengths and interests. Problem contexts often call on students to share experiences from their family, culture, or community. The curriculum eliminates barriers—such as work in isolation from peers, or emphasis on speed and memorization—that exclude some students from participating successfully in mathematics. The following aspects of the curriculum ensure that all students are included in significant mathematics learning:

- Students spend time exploring problems in depth.
- They find more than one solution to many of the problems they work on.
- They invent their own strategies and approaches, rather than rely on memorized procedures.
- They choose from a variety of concrete materials and appropriate technology, including calculators, as a natural part of their everyday mathematical work.
- They express their mathematical thinking through drawing, writing, and talking.
- They work in a variety of groupings—as a whole class, individually, in pairs, and in small groups.
- They move around the classroom as they explore the mathematics in their environment and talk with their peers.

While reading and other language activities are typically given a great deal of time and emphasis in elementary classrooms, mathematics often does not get the time it needs. If students are to experience mathematics in depth, they must have enough time to become engaged in real mathematical problems. We believe that a minimum of 5 hours of mathematics classroom time a week—about an hour a day—is critical at the elementary level. The scope and pacing of the *Investigations* curriculum are based on that belief.

We explain more about the pedagogy and principles that underlie these investigations in Teacher Notes throughout the units. For correlations of the curriculum to the NCTM Standards and further help in using this research-based program for teaching mathematics, see the following books, available from Dale Seymour Publications:

- *Implementing the* Investigations in Number, Data, and Space® *Curriculum*
- *Beyond Arithmetic: Changing Mathematics in the Elementary Classroom* by Jan Mokros, Susan Jo Russell, and Karen Economopoulos

This book is one of the curriculum units for *Investigations in Number, Data, and Space.* In addition to providing part of a complete mathematics curriculum for your students, this unit offers information to support your own professional development. You, the teacher, are the person who will make this curriculum come alive in the classroom; the book for each unit is your main support system.

Although the curriculum does not include student textbooks, reproducible sheets for student work are provided in the unit and are also available as Student Activity Booklets. Students work actively with objects and experiences in their own environment and with a variety of manipulative materials and technology, rather than with a book of instruction and problems. We strongly recommend use of the overhead projector as a way to present problems, to focus group discussion, and to help students share ideas and strategies.

Ultimately, every teacher will use these investigations in ways that make sense for his or her particular style, the particular group of students, and the constraints and supports of a particular school environment. Each unit offers information and guidance for a wide variety of situations, drawn from our collaborations with many teachers and students over many years. Our goal in this book is to help you, a professional educator, implement this curriculum in a way that will give all your students access to mathematical power.

Investigation Format

The opening two pages of each investigation help you get ready for the work that follows.

What Happens This gives a synopsis of each session or block of sessions.

Mathematical Emphasis This lists the most important ideas and processes students will encounter in this investigation.

What to Plan Ahead of Time These lists alert you to materials to gather, sheets to duplicate, transparencies to make, and anything else you need to do before starting.

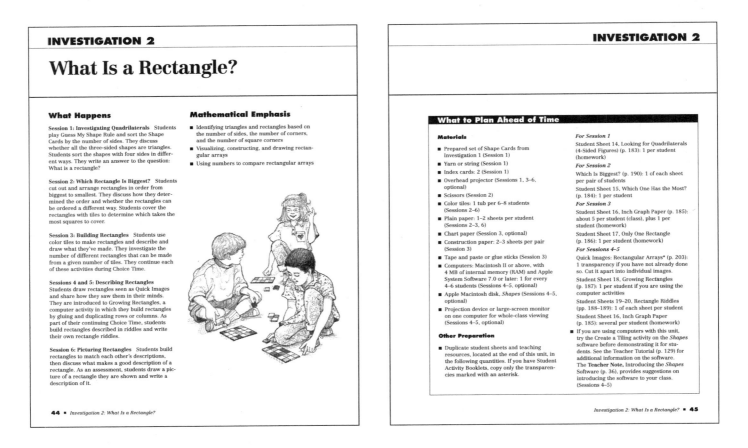

Sessions Within an investigation, the activities are organized by class session, a session being at least a one-hour math class. Sessions are numbered consecutively through an investigation. Often several sessions are grouped together, presenting a block of activities with a single major focus.

When you find a block of sessions presented together—for example, Sessions 1, 2, and 3—read through the entire block first to understand the overall flow and sequence of the activities. Make some preliminary decisions about how you will divide the activities into three sessions for your class, based on what you know about your students. You may need to modify your initial plans as you progress through the activities, and you may want to make notes in the margins of the pages as reminders for the next time you use the unit.

Be sure to read the Session Follow-Up section at the end of the session block to see what homework assignments and extensions are suggested as you make your initial plans.

While you may be used to a curriculum that tells you exactly what each class session should cover, we have found that the teacher is in a better position to make these decisions. Each unit is flexible and may be handled somewhat differently by every teacher. Although we provide guidance for how many sessions a particular group of activities is likely to need, we want you to be active in determining an appropriate pace and the best transition points for your class. It is not unusual for a teacher to spend more or less time than is proposed for the activities.

Classroom Routines The Start-Up at the beginning of each session offers suggestions for how to acknowledge and integrate homework from the previous session, and which Classroom Routine activities to include sometime during the school day. Routines provide students with regular practice in important mathematical skills such as solving number combinations, collecting and organizing data, understanding time, and seeing spatial relationships. Two routines, How Many Pockets? and Today's Number, are used regularly in the grade 2 *Investigations* units. A third routine, Time and Time Again, appears in the final unit, *Timelines and Rhythm Patterns*. This routine provides a variety of activities about understanding

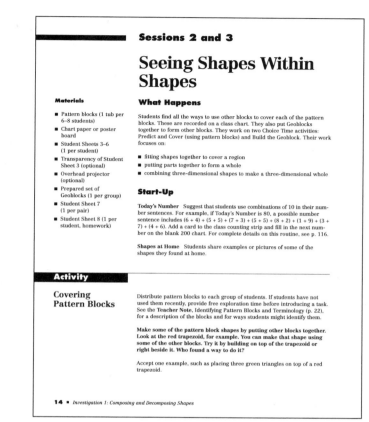

time; these can be easily integrated throughout the school day and into other parts of the classroom curriculum. A fourth routine, Quick Images, supports work in the unit *Shapes, Halves, and Symmetry*. After its introduction, you might do it once or twice a week to develop students' visual sense of number (as displayed in dot arrangements).

Most Classroom Routine activities are short and can be done whenever you have a spare 10 minutes—maybe before lunch or recess, or at the beginning or end of the day. Complete descriptions of the Classroom Routines can be found at the end of the units.

Activities The activities include pair and small-group work, individual tasks, and whole-class discussions. In any case, students are seated together, talking and sharing ideas during all work times. Students most often work cooperatively, although each student may record work individually.

Choice Time In most units, some sessions are structured with activity choices. In these cases, students may work simultaneously on different activities focused on the same mathematical ideas.

Students choose which activities they want to do, and they cycle through them.

You will need to decide how to set up and introduce these activities and how to let students make their choices. Some teachers set up choices as stations around the room, while others post the list of available choices and allow students to collect their own materials and choose their own work space. You may need to experiment with a few different structures before finding a set up that works best for you, your students, and your classroom.

Tips for the Linguistically Diverse Classroom At strategic points in each unit, you will find concrete suggestions for simple modifications of the teaching strategies to encourage the participation of all students. Many of these tips offer alternative ways to elicit critical thinking from students at varying levels of English proficiency, as well as from other students who find it difficult to verbalize their thinking.

The tips are supported by suggestions for specific vocabulary work to help ensure that all students can participate fully in the investigations. The Preview for the Linguistically Diverse Classroom lists important words that are assumed as part of the working vocabulary of the unit. Second-language learners will need to become familiar with these words in order to understand the problems and activities they will be doing. These terms can be incorporated into students' second-language work before or during the unit. Activities that can be used to present the words are found in the appendix, Vocabulary Support for Second-Language Learners. In addition, ideas for making connections to students' languages and cultures, included on the Preview page, help the class explore the unit's concepts from a multicultural perspective.

Session Follow-Up: Homework In *Investigations*, homework is an extension of classroom work. Sometimes it offers review and practice of work done in class, sometimes preparation for upcoming activities, and sometimes numerical practice that revisits work in earlier units. Homework plays a role both in supporting students' learning and in helping inform families about the ways in which students in this curriculum work with mathematical ideas.

Depending on your school's homework policies and your own judgment, you may want to assign more homework than is suggested in the units. For this purpose you might use the practice pages, included as blackline masters at the end of this unit, to give students additional work with numbers.

For some homework assignments, you will want to adapt the activity to meet the needs of a variety of students in your class: those with special needs, those ready for more challenge, and second-language learners. You might change the numbers in a problem, make the activity more or less complex, or go through a sample activity with those who need extra help. You can modify any student sheet for either homework or class use. In particular, making numbers in a problem smaller or larger can make the same basic activity appropriate for a wider range of students.

Another issue to consider is how to handle the homework that students bring back to class—how to recognize the work they have done at home without spending too much time on it. Some teachers hold a short group discussion of different approaches to the assignment; others ask students to share and discuss their work with a neighbor; still others post the homework around the room

and give students time to tour it briefly. If you want to keep track of homework students bring in, be sure it ends up in a designated place.

Session Follow-Up: Extensions Sometimes in Session Follow-Up, you will find suggested extension activities. These are opportunities for some or all students to explore a topic in greater depth or in a different context. They are not designed for "fast" students; mathematics is a multifaceted discipline, and different students will want to go further in different investigations. Look for and encourage the sparks of interest and enthusiasm you see in your students, and use the extensions to help them pursue these interests.

Excursions Some of the *Investigations* units include excursions—blocks of activities that could be omitted without harming the integrity of the unit. This is one way of dealing with the great depth and variety of elementary mathematics— much more than a class has time to explore in any one year. Excursions give you the flexibility to make different choices from year to year, doing the excursion in one unit this time, and next year trying another excursion.

Materials

A complete list of the materials needed for teaching this unit follows the unit overview. Some of these materials are available in kits for the *Investigations* curriculum. Individual items can also be purchased from school supply dealers.

Classroom Materials In an active mathematics classroom, certain basic materials should be available at all times: interlocking cubes, pencils, unlined paper, graph paper, calculators, and things to count with. Some activities in this curriculum require scissors and glue sticks or tape. Stick-on notes and large paper are also useful materials throughout.

So that students can independently get what they need at any time, they should know where these materials are kept, how they are stored, and how they are to be returned to the storage area. Many teachers have found that stopping 5 minutes before the end of each session so that students can finish their work and clean up is helpful in maintaining classroom materials. You'll find that establishing such routines at the beginning of the year is well worth the time and effort.

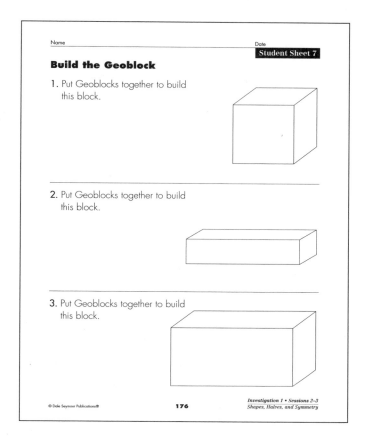

Student Sheets and Teaching Resources Student recording sheets and other teaching tools needed for both class and homework are provided as reproducible blackline masters at the end of each unit.

We think it's important that students find their own ways of organizing and recording their work. They need to learn how to explain their thinking with both drawings and written words, and how to organize their results so someone else can understand them. For this reason, we deliberately do not provide student sheets for every activity. Regardless of the form in which students do their work, we recommend that they keep their work in a mathematics folder, notebook, or journal so that it is always available to them for reference.

Student Activity Booklets These booklets contain all the sheets each student will need for individual work, freeing you from extensive copying (although you may need or want to copy the occasional teaching resource on transparency film or card stock, or make extra copies of a student sheet).

Calculators and Computers Calculators are introduced to students in the second unit of the grade 2 sequence, *Coins, Coupons, and Combinations.* It is assumed that calculators are readily available throughout the curriculum.

Computer activities are offered at all grade levels. Although the software is linked to activities in three units in grade 2, we recommend that students use it throughout the year. As students use the software over time, they continue to develop skills presented in the units. How you incorporate the computer activities into your curriculum depends on the number of computers you have available. Technology in the Curriculum discusses ways to incorporate the use of calculators and computers into classroom activities.

Children's Literature Each unit offers a list of related children's literature that can be used to support the mathematical ideas in the unit. Sometimes an activity is based on a selected children's book, with suggestions for substitutions where practical. While such activities can be adapted and taught without the book, the literature offers a rich introduction and should be used whenever possible.

Investigations at Home It is a good idea to make your policy on homework explicit to both students and their families when you begin teaching with *Investigations.* How frequently will you be assigning homework? When do you expect homework to be completed and brought back to school? What are your goals in assigning homework? How independent should families expect their children to be? What should the parent's or guardian's role be? The more explicit you can be about your expectations, the better the homework experience will be for everyone.

Investigations at Home (a booklet available separately for each unit, to send home with students) gives you a way to communicate with families about the work students are doing in class. This booklet includes a brief description of every session, a list of the mathematics content emphasized in each investigation, and a discussion of each homework assignment to help families more effectively support their children. Whether or not you are using the *Investigations* at Home booklets, we expect you to make your own choices about homework assignments. Feel free to omit any and to add extra ones you think are appropriate.

Family Letter A letter that you can send home to students' families is included with the blackline masters for each unit. Families need to be informed about the mathematics work in your classroom; they should be encouraged to participate in and support their children's work. A reminder to send home the letter for each unit appears in one of the early investigations. These letters are also available separately in Spanish, Vietnamese, Cantonese, Hmong, and Cambodian.

Help for You, the Teacher

Because we believe strongly that a new curriculum must help teachers think in new ways about mathematics and about their students' mathematical thinking processes, we have included a great deal of material to help you learn more about both.

About the Mathematics in This Unit This introductory section summarizes the critical informa-

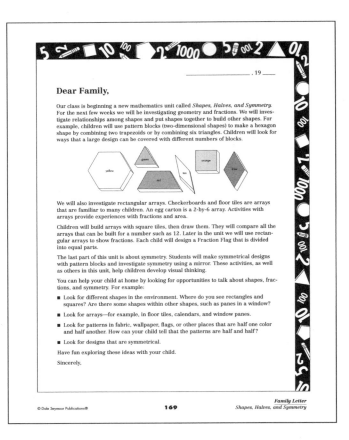

tion about the mathematics you will be teaching. It describes the unit's central mathematical ideas and the ways students will encounter them through the unit's activities.

About the Assessment in This Unit This introductory section highlights Teacher Checkpoints and assessment activities contained in the unit. It offers questions to stimulate your assessment as you observe the development of students' mathematical thinking and learning.

Teacher Notes These reference notes provide practical information about the mathematics you are teaching and about our experience with how students learn. Many of the notes were written in response to actual questions from teachers or to discuss important things we saw happening in the field-test classrooms. Some teachers like to read them all before starting the unit, then review them as they come up in particular investigations.

Dialogue Boxes Sample dialogues demonstrate how students typically express their mathematical ideas, what issues and confusions arise in their thinking, and how some teachers have guided class discussions.

These dialogues are based on the extensive classroom testing of this curriculum; many are word-for-word transcriptions of recorded class discussions. They are not always easy reading; sometimes it may take some effort to unravel what the students are trying to say. But this is the value of these dialogues; they offer good clues to how your students may develop and express their approaches and strategies, helping you prepare for your own class discussions.

Where to Start You may not have time to read everything the first time you use this unit. As a first-time user, you will likely focus on understanding the activities and working them out with your students. Read completely through all the activities before starting to present them. Also read those sections listed in the Contents under the heading Where to Start.

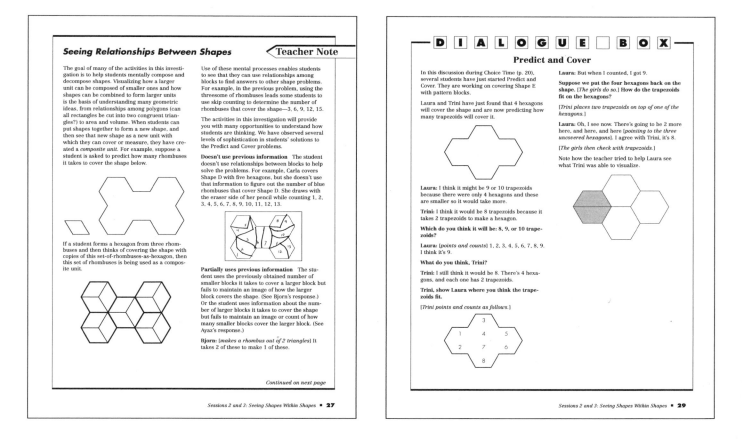

The *Investigations* curriculum incorporates the use of two forms of technology in the classroom: calculators and computers. Calculators are assumed to be standard classroom materials, available for student use in any unit. Computers are explicitly linked to one or more units at each grade level; they are used with the unit on 2-D geometry at each grade, as well as with some of the units on measuring, data, and changes.

Using Calculators

In this curriculum, calculators are considered tools for doing mathematics, similar to pattern blocks or interlocking cubes. Just as with other tools, students must learn both *how* to use calculators correctly and *when* they are appropriate to use. This knowledge is crucial for daily life, as calculators are now a standard way of handling numerical operations, both at work and at home.

Using a calculator correctly is not a simple task; it depends on a good knowledge of the four operations and of the number system, so that students can select suitable calculations and also determine what a reasonable result would be. These skills are the basis of any work with numbers, whether or not a calculator is involved.

Unfortunately, calculators are often seen as tools to check computations with, as if other methods are somehow more fallible. Students need to understand that any computational method can be used to check any other; it's just as easy to make a mistake on the calculator as it is to make a mistake on paper or with mental arithmetic. Throughout this curriculum, we encourage students to solve computation problems in more than one way in order to double-check their accuracy. We present mental arithmetic, paper-and-pencil computation, and calculators as three possible approaches.

In this curriculum we also recognize that, despite their importance, calculators are not always appropriate in mathematics instruction. Like any tools, calculators are useful for some tasks but not for others. You will need to make decisions about when to allow students access to calculators and when to ask that they solve problems without

them so that they can concentrate on other tools and skills. At times when calculators are or are not appropriate for a particular activity, we make specific recommendations. Help your students develop their own sense of which problems they can tackle with their own reasoning and which ones might be better solved with a combination of their own reasoning and the calculator.

Managing calculators in your classroom so that they are a tool, and not a distraction, requires some planning. When calculators are first introduced, students often want to use them for everything, even problems that can be solved quite simply by other methods. However, once the novelty wears off, students are just as interested in developing their own strategies, especially when these strategies are emphasized and valued in the classroom. Over time, students will come to recognize the ease and value of solving problems mentally, with paper and pencil, or with manipulatives, while also understanding the power of the calculator to facilitate work with larger numbers.

Experience shows that if calculators are available only occasionally, students become excited and distracted when they are permitted to use them. They focus on the tool rather than on the mathematics. In order to learn when calculators are appropriate and when they are not, students must have easy access to them and use them routinely in their work.

If you have a calculator for each student, and if you think your students can accept the responsibility, you might allow them to keep their calculators with the rest of their individual materials, at least for the first few weeks of school. Alternatively, you might store them in boxes on a shelf, number each calculator, and assign a corresponding number to each student. This system can give students a sense of ownership while also helping you keep track of the calculators.

Using Computers

Students can use computers to approach and visualize mathematical situations in new ways. The computer allows students to construct and manipulate geometric shapes, see objects move according to rules they specify, and turn, flip, and repeat a pattern.

This curriculum calls for computers in units where they are a particularly effective tool for learning mathematics content. One unit on 2-D geometry at each of the grades 3–5 includes a core of activities that rely on access to computers, either in the classroom or in a lab. Other units on geometry, measuring, data, and changes include computer activities, but can be taught without them. In these units, however, students' experience is greatly enhanced by computer use.

The following list outlines the recommended use of computers in this curriculum:

Kindergarten
Unit: *Making Shapes and Building Blocks* (Exploring Geometry)
Software: *Shapes*
Source: provided with the unit

Grade 1
Unit: *Survey Questions and Secret Rules* (Collecting and Sorting Data)
Software: *Tabletop, Jr.*
Source: Broderbund

Unit: *Quilt Squares and Block Towns* (2-D and 3-D Geometry)
Software: *Shapes*
Source: provided with the unit

Grade 2
Unit: *Mathematical Thinking at Grade 2* (Introduction)
Software: *Shapes*
Source: provided with the unit

Unit: *Shapes, Halves, and Symmetry* (Geometry and Fractions)
Software: *Shapes*
Source: provided with the unit

Unit: *How Long? How Far?* (Measuring)
Software: *Geo-Logo*
Source: provided with the unit

Grade 3
Unit: *Flips, Turns, and Area* (2-D Geometry)
Software: *Tumbling Tetrominoes*
Source: provided with the unit

Unit: *Turtle Paths* (2-D Geometry)
Software: *Geo-Logo*
Source: provided with the unit

Grade 4
Unit: *Sunken Ships and Grid Patterns* (2-D Geometry)
Software: *Geo-Logo*
Source: provided with the unit

Grade 5
Unit: *Picturing Polygons* (2-D Geometry)
Software: *Geo-Logo*
Source: provided with the unit

Unit: *Patterns of Change* (Tables and Graphs)
Software: *Trips*
Source: provided with the unit

Unit: *Data: Kids, Cats, and Ads* (Statistics)
Software: *Tabletop, Sr.*
Source: Broderbund

The software provided with the *Investigations* units uses the power of the computer to help students explore mathematical ideas and relationships that cannot be explored in the same way with physical materials. With the *Shapes* (grades 1–2) and *Tumbling Tetrominoes* (grade 3) software, students explore symmetry, pattern, rotation and reflection, area, and characteristics of 2-D shapes. With the *Geo-Logo* software (grades 2–5), students investigate rotations and reflections, coordinate geometry, the properties of 2-D shapes, and angles. The *Trips* software (grade 5) is a mathematical exploration of motion in which students run experiments and interpret data presented in graphs and tables.

We suggest that students work in pairs on the computer; this not only maximizes computer resources but also encourages students to consult, monitor, and teach each other. Generally, more than two students at one computer find it difficult to share. Managing access to computers is an issue for every classroom. The curriculum gives you explicit support for setting up a system. The units are structured on the assumption that you have enough computers for half your students to work on the machines in pairs at one time. If you do not have access to that many computers, suggestions are made for structuring class time to use the unit with fewer than five.

Assessment plays a critical role in teaching and learning, and it is an integral part of the *Investigations* curriculum. For a teacher using these units, assessment is an ongoing process. You observe students' discussions and explanations of their strategies on a daily basis and examine their work as it evolves. While students are busy recording and representing their work, working on projects, sharing with partners, and playing mathematical games, you have many opportunities to observe their mathematical thinking. What you learn through observation guides your decisions about how to proceed. In any of the units, you will repeatedly consider questions like these:

- Do students come up with their own strategies for solving problems, or do they expect others to tell them what to do? What do their strategies reveal about their mathematical understanding?

- Do students understand that there are different strategies for solving problems? Do they articulate their strategies and try to understand other students' strategies?

- How effectively do students use materials as tools to help with their mathematical work?

- Do students have effective ideas for keeping track of and recording their work? Do keeping track of and recording their work seem difficult for them?

You will need to develop a comfortable and efficient system for recording and keeping track of your observations. Some teachers keep a clipboard handy and jot notes on a class list or on adhesive labels that are later transferred to student files. Others keep loose-leaf notebooks with a page for each student and make weekly notes about what they have observed in class.

Assessment Tools in the Unit

With the activities in each unit, you will find questions to guide your thinking while observing the students at work. You will also find two built-in assessment tools: Teacher Checkpoints and embedded Assessment activities.

Teacher Checkpoints The designated Teacher Checkpoints in each unit offer a time to "check in" with individual students, watch them at work, and ask questions that illuminate how they are thinking.

At first it may be hard to know what to look for, hard to know what kinds of questions to ask. Students may be reluctant to talk; they may not be accustomed to having the teacher ask them about their work, or they may not know how to explain their thinking. Two important ingredients of this process are asking students open-ended questions about their work and showing genuine interest in how they are approaching the task. When students see that you are interested in their thinking and are counting on them to come up with their own ways of solving problems, they may surprise you with the depth of their understanding.

Teacher Checkpoints also give you the chance to pause in the teaching sequence and reflect on how your class is doing overall. Think about whether you need to adjust your pacing: Are most students fluent with strategies for solving a particular kind of problem? Are they just starting to formulate good strategies? Or are they still struggling with how to start? Depending on what you see as the students work, you may want to spend more time on similar problems, change some of the problems to use smaller numbers, move quickly to more challenging material, modify subsequent activities for some students, work on particular ideas with a small group, or pair students who have good strategies with those who are having more difficulty.

Embedded Assessment Activities Assessment activities embedded in each unit will help you examine specific pieces of student work, figure out what they mean, and provide feedback. From the students' point of view, these assessment activities are no different from any others. Each is a learning experience in and of itself, as well as an opportunity for you to gather evidence about students' mathematical understanding.

The embedded assessment activities sometimes involve writing and reflecting; at other times, a discussion or brief interaction between student and teacher; and in still other instances, the creation and explanation of a product. In most cases, the assessments require that students *show* what they did, *write* or *talk* about it, or do both. Having to explain how they worked through a problem helps students be more focused and clear in their mathematical thinking. It also helps them realize that doing mathematics is a process that may involve tentative starts, revising one's approach, taking different paths, and working through ideas.

Teachers often find the hardest part of assessment to be interpreting their students' work. We provide guidelines to help with that interpretation. If you have used a process approach to teaching writing, the assessment in *Investigations* will seem familiar. For many of the assessment activities, a Teacher Note provides examples of student work and a commentary on what it indicates about student thinking.

Documentation of Student Growth

To form an overall picture of mathematical progress, it is important to document each student's work. Many teachers have students keep their work in folders, notebooks, or journals, and some like to have students summarize their learning in journals at the end of each unit. It's important to document students' progress, and we recommend that you keep a portfolio of selected work for each student, unit by unit, for the entire year. The final activity in each *Investigations* unit, called Choosing Student Work to Save, helps you and the students select representative samples for a record of their work.

This kind of regular documentation helps you synthesize information about each student as a mathematical learner. From different pieces of evidence, you can put together the big picture. This synthesis will be invaluable in thinking about where to go next with a particular child, deciding where more work is needed, or explaining to parents (or other teachers) how a child is doing.

If you use portfolios, you need to collect a good balance of work, yet avoid being swamped with an overwhelming amount of paper. Following are some tips for effective portfolios:

- Collect a representative sample of work, including some pieces that students themselves select for inclusion in the portfolio. There should be just a few pieces for each unit, showing different kinds of work—some assignments that involve writing as well as some that do not.

- If students do not date their work, do so yourself so that you can reconstruct the order in which pieces were done.

- Include your reflections on the work. When you are looking back over the whole year, such comments are reminders of what seemed especially interesting about a particular piece; they can also be helpful to other teachers and to parents. Older students should be encouraged to write their own reflections about their work.

Assessment Overview

There are two places to turn for a preview of the assessment opportunities in each *Investigations* unit. The Assessment Resources column in the unit Overview Chart identifies the Teacher Checkpoints and Assessment activities embedded in each investigation, guidelines for observing the students that appear within classroom activities, and any Teacher Notes and Dialogue Boxes that explain what to look for and what types of student responses you might expect to see in your classroom. Additionally, the section About the Assessment in This Unit gives you a detailed list of questions for each investigation, keyed to the mathematical emphases, to help you observe student growth.

Depending on your situation, you may want to provide additional assessment opportunities. Most of the investigations lend themselves to more frequent assessment, simply by having students do more writing and recording while they are working.

Shapes, Halves, and Symmetry

Content of This Unit Using pattern blocks, Geoblocks, square tiles, and the *Shapes* computer software, students explore the structure of shapes and how they can be decomposed or put together into other shapes. They investigate the structure of rectangular arrays by covering rectangles with tiles, and by building, drawing, and describing rectangles. They find halves of rectangles and other two- and three-dimensional shapes. They explore symmetry by making symmetrical designs and pictures.

Connections with Other Units If you are doing the full-year *Investigations* curriculum in the grade 2 sequence, this is the fourth of eight units. In the introductory unit, *Mathematical Thinking at Grade 2*, students began their investigation of two- and three-dimensional shapes. Students will continue their study of geometry and measurement in the unit *How Long? How Far?*

This unit also can be used successfully at grade 3, depending on the previous experience and needs of your students.

Investigations Curriculum ■ Suggested Grade 2 Sequence

Mathematical Thinking at Grade 2 (Introduction)

Coins, Coupons, and Combinations (The Number System)

Does It Walk, Crawl, or Swim? (Sorting and Classifying Data)

▶ *Shapes, Halves, and Symmetry* (Geometry and Fractions)

Putting Together and Taking Apart (Addition and Subtraction)

How Long? How Far? (Measuring)

How Many Pockets? How Many Teeth? (Collecting and Representing Data)

Timelines and Rhythm Patterns (Representing Time)

Investigation 1 ▪ Composing and Decomposing Shapes

Class Sessions	Activities	Pacing
Session 1 (p. 5) SHAPES AROUND US	Quick Images: A New Routine Shapes in the Classroom Sorting Shape Cards Making Shape Card Posters Introducing Math Folders and Weekly Logs Homework: Shapes at Home	minimum 1 hr
Sessions 2 and 3 (p. 14) SEEING SHAPES WITHIN SHAPES	Covering Pattern Blocks Predict and Cover Build the Geoblock Introducing Choice Time Homework: Shapes Within Shapes	minimum 2 hr
Sessions 4 and 5 (p. 30) SHAPE PUZZLES	On-Computer Activity: Introducing Solve Puzzles Choice Time Teacher Checkpoint: Predict and Cover Homework: Composing New Shapes with 2 Triangles	minimum 2 hr
Sessions 6, 7, and 8 (p. 38) BUILDING BUILDINGS	The Last Block Game Cube Buildings Choice Time Class Discussion: Build a Building Homework: Composing New Shapes with 3 or 4 Triangles Extension: Guess My Building	minimum 3 hr

Start-Up ▪ Today's Number, Quick Images

Mathematical Emphasis

- Sorting, describing, and identifying shapes by various attributes

- Composing and decomposing two- and three-dimensional shapes

- Describing spatial and numerical relationships found among shapes

Assessment Resources

Observing the Students (p. 8)

Categorizing Shape Cards (Dialogue Box, p. 12)

Observing the Students (p. 15)

Observing the Students (p. 19)

Keeping Track of Students' Work (Teacher Note, p. 26)

Seeing Relationships Between Shapes (Teacher Note, p. 27)

Predict and Cover (Dialogue Box, p. 29)

Observing the Students (p. 32)

Teacher Checkpoint: Predict and Cover (p. 33)

Observing the Students (p. 41)

Students' Thinking About Build a Building (Teacher Note, p. 43)

Materials

Student math folder

Paste or glue stick

Scissors

Large paper (18" by 24")

Overhead projector

Plastic bags or envelopes

Chart paper

Pattern blocks

Geoblocks

Large-screen monitor or projection device

Computers

Shapes Software

Interlocking cubes

Transparent pattern blocks

Student Sheets 1–13

Family letter

Teaching resource sheets

Investigation 2 ■ What Is a Rectangle?

Class Sessions	Activities	Pacing
Session 1 (p. 46) INVESTIGATING QUADRILATERALS	Guess My Shape Rule Writing: What Is a Rectangle? Homework: Looking for Quadrilaterals (4-Sided Figures)	minimum 1 hr
Session 2 (p. 52) WHICH RECTANGLE IS BIGGEST?	Ordering Rectangles Covering Rectangles	minimum 1 hr
Session 3 (p. 55) BUILDING RECTANGLES	Building Tile Rectangles How Many Rectangles? Choice Time Homework: Only One Rectangle	minimum 1 hr
Sessions 4 and 5 (p. 62) DESCRIBING RECTANGLES	Quick Images: Rectangular Arrays On-Computer Activity: Introducing Growing Rectangles Choice Time Homework: Making Rectangles Extension: Fill the Rectangles	minimum 2 hr
Session 6 (p. 66) PICTURING RECTANGLES	Describing Rectangles Assessment: Picturing a Rectangle	minimum 1 hr

Start-Up ■ Quick Images, Today's Number

Mathematical Emphasis

- Identifying triangles and rectangles based on the number of sides, the number of corners, and the number of square corners

- Visualizing, constructing, and drawing rectangular arrays

- Using numbers to compare rectangular arrays

Assessment Resources

What's a Rectangle?: (Teacher Note, p. 50)

Choice Time: Observing the Students (p. 58)

Is It the Same Rectangle? (Dialogue Box, p. 60)

Strategies for Building Rectangles (Dialogue Box, p. 61)

Choice Time: Observing the Students (p. 64)

Assessment: Picturing a Rectangle (p. 68)

Assessment: Picturing Rectangles (Teacher Note, p. 69)

Materials

Yarn or string

Index cards

Overhead projector

Scissors

Color tiles

Plain paper

Chart paper

Construction paper

Tape

Paste or glue stick

Computers

Shapes software

Large-screen monitor or projection device

Student Sheets 14–20

Teaching resource sheets

Investigation 3 ▪ Fractions of Geometric Shapes

Class Sessions	Activities	Pacing
Sessions 1 and 2 (p. 74) HALVES OF RECTANGLES AND SOLIDS	Half-and-Half Rectangles Introducing Halves of Geoblocks Teacher Checkpoint: Choice Time Teacher Checkpoint: Halves Homework: Half-and-Half Rectangles Homework: Things That Come in Halves	minimum 2 hr
Sessions 3, 4, and 5 (p. 82) CUTTING CONGRUENT HALVES	Introducing Shape Halves Choice Time Class Discussion: Which Rectangles Make Halves? Homework: Designing Shapes That Can Be Cut in Half	minimum 3 hr
Session 6 (p. 86) FRACTION FLAGS	Fraction Flags Homework: Half-and-Half Flags Extension: Congruent-Halves Flags Extension: Halves Display	minimum 1 hr
Sessions 7 and 8 (Excursion) (p. 89)* FOURTHS AND THIRDS OF RECTANGLES	Fourths and Thirds of Rectangles Thirds and Fourths Flags Extension: Thirds and Fourths Displays	minimum 2 hr

Start-Up ▪ Today's Number, Quick Images

*Excursions can be omitted without harming the integrity or continuity of the unit, but offer good mathematical work if you have time to include them.

Mathematical Emphasis

- Constructing arrays to represent numbers and identifying halves of the arrays

- Investigating halves of three-dimensional solids

- Constructing two-dimensional arrays that are divided into thirds and fourths

- Describing fractional parts of an array as fractions of a rectangular region

- Describing fractional parts of an array as fractions of the set of tiles used to construct the array

- Designing and constructing a rect-angular region that is divided into halves, thirds, or fourths

Assessment Resources

Teacher Checkpoint: Halves (p. 78)

Halves of Rectangles (Teacher Note, p. 80)

What Is a Half? (Dialogue Box, p. 81)

Choice Time: Observing the Students (p. 84)

Fourths and Thirds of Rectangles: Observing the Students (p. 92)

Materials

Geoblocks
Color tiles
Crayons or markers
Chart paper
Scissors
Overhead projector
Construction paper
Plastic bags or envelopes
Paste or glue sticks
Pictures of flags
Drawing paper
Transparent color tiles
Student Sheets 21–24
Teaching resource sheets

Investigation 4 ▪ Symmetry

Class Sessions	Activities	Pacing
Sessions 1 and 2 (p. 96) SYMMETRICAL DESIGNS	Symmetry in the World Introducing Geoblock Buildings Choice Time Homework: Looking for Symmetry	minimum 2 hr
Sessions 3 and 4 (p. 104) REFLECTING BLOCKS AND TILES	Mirror Designs Introducing Copy Tiles Choice Time Homework: Exploring Mirror Symmetry	minimum 2 hr
Sessions 5 and 6 (p. 109) PAPER FOLDING AND CUTTING	Introducing Fold and Cut Choice Time Class Discussion: Is It Symmetrical?	minimum 2 hr
Session 7 (p. 113) SYMMETRICAL PICTURES	Assessment: Symmetrical Pictures Choosing Student Work to Save Homework: Fold and Cut	minimum 1 hr

Start-Up ▪ Today's Number, Quick Images

Mathematical Emphasis

- Finding and describing objects that have mirror symmetry

- Making two-dimensional symmetrical designs

- Building three-dimensional symmetrical structures

Assessment Resources

Choice Time: Observing the Students (p. 100)

Discussing Symmetry (Dialogue Box, p. 103)

Choice Time: Observing the Students (p. 107)

Choice Time: Observing the Students (p. 111)

Assessment: Symmetrical Pictures (p. 113)

Choosing Student Work to Save (p. 115)

Materials

Pattern block stickers

Transparent pattern blocks

Pattern blocks

Geoblocks

Computers

Shapes software

Large-screen monitor or projection device

Crayons or markers

Overhead projector

Mirrors

Scissors

Color tiles

Construction paper

Chart paper

Plain paper

Letter-size paper

Paste or glue sticks

Student Sheets 25–27

Teaching resource sheets

Following are the basic materials needed for the activities in this unit. Many of the items can be purchased from the publisher, either individually or in the Teacher Resource Package and the Student Materials Kit for grade 2. Detailed information is available on the *Investigations* order form. To obtain this form, call toll-free 1-800-872-1100 and ask for a Dale Seymour customer service representative.

Geoblocks (2 sets, each divided into 2–3 subsets)

Snap™ Cubes (interlocking cubes) (about 30 per student)

Wooden pattern blocks (1 tub per 6–8 students)

Overhead pattern blocks (optional)

Pattern block stickers (optional)

Square color tiles (1 tub per 6–8 students)

Overhead color tiles (optional)

Mirrors (1 per 3–4 students)

Apples Macintosh disk, *Shapes*

Computers: Macintosh II or above, with 4 MB of internal memory (RAM) and Apple System Software 7.0 or later (1 for every 4–6 students, optional)

Projection device or large screen monitor on one computer for whole class viewing (optional)

Overhead projector (optional)

Student math folders (1 per student)

Resealable plastic bags or envelopes (about 48)

Chart paper

Large paper, 18" by 24" (about 75 sheets)

Letter-size paper in half sheets (about 6 per student)

Drawing paper (about 150 sheets)

Index cards (2)

Construction paper (about 300 sheets)

Crayons or markers

Scissors

Paste or glue sticks

Tape

Large jar

Yarn or string

Pictures of flags (especially those that show halves, optional)

The following materials are provided at the end of this unit as blackline masters. A Student Activity Booklet containing all student sheets and teaching resources needed for individual work is available.

Family Letter (p. 169)

Student Sheets 1–27 (p. 170)

Teaching Resources:

   Which Is Biggest? (p. 190)

   Shape Halves (p. 196)

   Quick Images (p. 201)

   Shape Cards (p. 205)

Practice Pages (p. 209)

Related Children's Literature

Burns, Marilyn. *The Greedy Triangle*. New York: Scholastic Inc., 1994.

Emberley, Ed. *Ed Emberley's Picture Pie: A Circle Drawing Book*. Boston: Little, Brown and Company, 1984.

Hoban, Tana. *Shapes, Shapes, Shapes*. New York: Greenwillow Books, 1986.

Hutchins, Pat. *The Doorbell Rang*. New York: Greenwillow Books, 1986.

McMillan, Bruce. *Eating Fractions*. New York: Scholastic Inc., 1991.

Pomerantz, Charlotte. *The Half Birthday Party*. New York: Clarion Books, 1984.

Testa, Fulvio. *If You Look Around You*. New York: Dial Books for Young Readers, 1983.

One of the major ideas in this unit is that shapes can be combined or decomposed to make other shapes. As students use pattern blocks or Geoblocks, they explore relationships within each of these sets. For example, they notice that three pattern block triangles can be combined to make a trapezoid or that two cubes can be combined to make a rectangular solid, using the Geoblocks. Just as students compose and decompose numbers to make them more manageable when they are solving numerical problems, solving geometric problems often involves understanding how to pull shapes apart and put them together, so that they can be viewed in new ways. Much of what we understand about geometric shapes is based on our knowledge of how shapes are related. When second graders see how a regular hexagon can be constructed from six equilateral triangles, they are developing a foundation that will help them investigate the properties of the hexagon's angles, sides, and area.

As students explore two-dimensional shapes, they also begin to think about the similarities and differences of those shapes. This focus leads to classification and definition of basic shapes. Classification and definition are important processes in geometry: What makes a triangle a triangle? Why is a square considered to be a rectangle? What are the characteristics of a cube? Classification is an important way of organizing the world for yourself and is central to young students' learning. As students work to define what a rectangle is, they develop ideas by examining and comparing many examples of shapes. They think about which properties are important and which are not. Is a rectangle still a rectangle if it is oriented so that one of its corners is pointed toward the bottom of the page? Is it still a rectangle if it is a rectangular shape with a "bite" taken out of one corner? In developing definitions, students are learning to describe geometric characteristics and to look carefully at similarities and differences among shapes.

Students' study of shape leads them to consider basic ideas about area and volume. As students create rectangles with square tiles or create rectangular solids with cubes, they see how surfaces can be covered or space can be filled with many of the same unit. Here, their explorations in geometry are closely linked with number. Determining the number of tiles used to make a rectangle or the number of cubes used to make a rectangular solid leads students to look at the structure of rows or layers. They use a variety of numerical strategies, based on addition and multiplication, to calculate the total number of cubes or tiles.

Another connection between shape and number is encountered in another major focus of the unit—fractions, particularly halves, of shapes. This emphasis is closely related to the work on composing and decomposing shapes. When shapes are decomposed into equal parts, these parts are fractions of the whole shape. Students also encounter the idea of congruent parts of a region, that is, parts that are exactly the same size and shape.

The unit ends with explorations of symmetry. While many students intuitively recognize that many things are symmetrical, they need the opportunity to explore symmetry and think about it in more depth. As they create or investigate symmetric shapes, they develop clearer language and ideas about what symmetry is and how it behaves.

At the beginning of each investigation, the Mathematical Emphasis section tells you what is most important for students to learn about during that investigation. Many of these mathematical understandings and processes are difficult and complex. Students gradually learn more and more about each idea over many years of schooling. Individual students will begin and end the unit with different levels of knowledge and skill, but all will gain greater knowledge of geometric shapes, their relationships, and their properties.

Throughout the *Investigations* curriculum, there are many opportunities for ongoing daily assessment as you observe, listen to, and interact with students at work. In this unit, you will find three Teacher Checkpoints:

Investigation 1, Sessions 4–5: Predict and Cover (p. 33)

Investigation 3, Sessions 1–2: Choice Time (p. 77)

Investigation 3, Sessions 1–2: Halves (p. 78)

This unit also has two embedded assessment activities:

Investigation 2, Session 6: Picturing a Rectangle (p. 68)

Investigation 4, Session 7: Symmetrical Pictures (p. 113)

In addition, you can use almost any activity in this unit to assess your students' needs and strengths. Listed below are questions to help you focus your observations in each investigation. You may want to keep track of your observations for each student to help you plan your curriculum and monitor students' growth. Suggestions for documenting student growth can be found in the section About Assessment.

Investigation 1: Composing and Decomposing Shapes

- How do students sort, describe, and identify shapes? Which attributes do they attend to? Do students handle shapes that fit in more than one of the categories they've created? How?

- How flexibly do students compose and decompose shapes? Do they seem to have a plan? Do they use trial and error? How familiar are students with the shapes themselves? Do they see relationships between and among the shapes? Do they use these relationships to compose and decompose larger shapes? How?

- Do students describe and use the spatial and numerical relationships among shapes? How? Do some students seem to view shapes in primarily spatial or numerical terms? For example, do students see that 3 blue rhombuses take up the same space as 6 green triangles? Or do they focus on the numerical relationship between 3 and 6?

Investigation 2: What Is a Rectangle?

- How do students identify and define triangles and rectangles? What shapes do they include? Do they include all types of triangles and rectangles? Do they attend to the number of sides? to the number or corners? to the type or corner (for example, square)?

- How do students visualize a rectangular array based on a description? Can they build the array? What features do students attend to as they describe a rectangular array? How do students draw an array? Can they draw from a picture in their mind? from a constructed array? Are the shapes they construct rectangles?

- Do students compare rectangular arrays? How? Do they describe rectangular arrays in terms of the number in each row and column?

Investigation 3: Fractions of Geometric Shapes

- How do students represent numbers with arrays? Can they find more than one array for a given number? Do students identify half of the array? How? Can they create arrays that are half one color and half another? Can they keep track of the total number while thinking about halves?

- Do students find halves of three-dimensional solids? How? Do students find more than one block that is half of another? Do students recognize that all the half-blocks are the same size?

- How do students divide two-dimensional arrays into thirds and fourths? Do they identify numbers that make rectangles that can (and cannot) be divided into halves, thirds, and fourths? Do they see patterns in these numbers?

- Do students describe fractional parts of an array? How? Do they describe a region as being a certain fraction of the whole array? ("One fourth of my array is blue.") Do they describe the fraction numerically by saying how many tiles of the total are in that part? ("Three of the 12 tiles are blue.")

- Do students design and construct a rectangle that is divided into halves, thirds, or fourths? How? Do they decide on a total number of tiles and then divide the tiles up? Do they take a certain number for each fractional part and combine the numbers for the total?

Investigation 4: Symmetry

- Can students identify objects in the environment that have mirror symmetry? How do they describe mirror symmetry? How do they explain why something is or is not symmetrical?

- Do students make two- and three-dimensional symmetrical designs and structures? How? Are students' creations symmetrical? Can they identify a design that is not symmetrical? Can they fix it? How elaborate are their designs? Do they have more than one line of symmetry? Can they identify the lines of symmetry?

In the *Investigations* curriculum, mathematical vocabulary is introduced naturally during the activities. We don't ask students to learn definitions of new terms; rather, they come to understand such words as *factor* or *area* or *symmetry* by hearing them used frequently in discussion as they investigate new concepts. This approach is compatible with current theories of second-language acquisition, which emphasize the use of new vocabulary in meaningful contexts while students are actively involved with objects, pictures, and physical movement.

Listed below are some key words used in this unit that will not be new to most English speakers at this age level but may be unfamiliar to students with limited English proficiency. You will want to spend additional time working on these words with your students who are learning English. If you are working with a second-language teacher, you might enlist your colleague's aid in familiarizing students with these words before and during this unit. In the classroom, look for opportunities for students to hear and use these words. Activities you can use to present the words are given in the appendix, Vocabulary Support for Second-Language Learners (p. 127).

shape, fewest Throughout the unit, students explore the structure of two- and three-dimensional shapes, sometimes looking for ways to cover two-dimensional shapes with the fewest number of blocks.

predict, cover Students use relationships they've discovered among pattern blocks to make predictions about how many blocks it takes to cover shapes.

design, match Students build pattern block designs and look for blocks that match given shapes.

stories, floor, buildings Students use interlocking cubes to build and count the number of "rooms" in buildings with multiple floors or stories.

Multicultural Extensions for All Students

Whenever possible, encourage students to share words, objects, customs, or any aspects of daily life from their own cultures and backgrounds that are relevant to the activities in this unit. For example:

- Post photographs of types of buildings and structures found in cities around the world as students are building their structures during Investigation 1. You might discuss the shapes (such as square, round, rectangular) or sizes of these buildings.

- As students work on the Fraction Flag activity during Investigation 3, encourage them to share designs of flags from countries around the world with which they are familiar. Ask them to think about whether the flag has equivalent parts (for example, the Irish and Italian flags). Students can draw these flags and share their pictures, describing the fractional parts.

- While students are completing the folding and cutting activity in Investigation 4, tell them about the Japanese art of origami, which involves folding thin sheets of paper into shapes of animals, birds, and objects. You may want to try a simple folding activity with students, which you can find in a book about origami. Much of the folding will create symmetry along the fold lines, and many of the finished objects will be symmetrical. Students may be able to identify triangles and trapezoids in an origami crane, for instance, and recognize that the finished crane is symmetrical.

Investigations

Composing and Decomposing Shapes

What Happens

Session 1: Shapes Around Us   Students look for two-dimensional shapes in the classroom. They describe shapes from a set of Shape Cards and place together those with the same attribute. They sort the cards in different ways and make a poster to show one way of sorting.

Sessions 2 and 3: Seeing Shapes Within Shapes Students find all the ways to use other blocks to cover each of the pattern blocks. These are recorded on a class chart. They also put Geoblocks together to form other blocks. They work on two Choice Time activities: Predict and Cover (using pattern blocks) and Build the Geoblock.

Sessions 4 and 5: Shape Puzzles   Students are introduced to a new Choice Time activity, Solve Puzzles (using the *Shapes*™ software), and continue to work on the activities introduced previously. In the last half of Session 5, students share their results for Predict and Cover and look for patterns that help them make predictions.

Sessions 6, 7, and 8: Building Buildings Students are introduced to two additional Choice Time activities. In The Last Block Game, students take turns covering shapes with pattern blocks. In Build a Building, they use interlocking cubes to build and count the number of "rooms" in multistory buildings. Students also continue work on other Choice Time activities. At the end of Session 8, they look for numerical patterns in their buildings.

Mathematical Emphasis

- Sorting, describing, and identifying shapes by various attributes
- Composing and decomposing two- and three-dimensional shapes
- Describing spatial and numerical relationships found among shapes

What to Plan Ahead of Time

Materials

- Student math folder: 1 per student (Session 1)
- Paste or glue sticks (Session 1)
- Scissors (Session 1)
- Large paper, 18" by 24" or larger: about 15 sheets (Session 1)
- Overhead projector (Sessions 1–8, optional)
- Resealable plastic bags or envelopes for storage of prepared sets of Quick Images, Shape Cards, and interlocking cubes: about 25 (Sessions 1, 6–8)
- Chart paper or poster board (Sessions 2–3)
- Pattern blocks: 1 tub per 6–8 students (Sessions 2–8)
- Geoblocks: 2 sets, each divided into 2 or 3 subsets as described in Other Preparation (p. 4) (Sessions 2–8)
- Projection device or large-screen monitor on one computer for whole-class viewing (Sessions 4–5, optional)
- Computers: Macintosh II or above, with 4 MB of internal memory (RAM) and Apple System Software 7.0 or later: 1 for every 4–6 students (Sessions 4–8, optional)
- Apple Macintosh disk, *Shapes* (Sessions 4–8, optional)
- Interlocking cubes: 60 per pair, stored in resealable plastic bags or plastic tubs with lids (Sessions 6–8)
- Transparent pattern blocks (Sessions 6–8, optional)

Other Preparation

- Duplicate student sheets and teaching resources (located at the end of this unit) in the following quantities. If you have Student Activity Booklets, copy only the items marked with an asterisk.

For Session 1

Student Sheet 1, Weekly Log (p. 170): 1 per student. At this time, you may wish to duplicate a supply to last for the entire unit and distribute the sheets as needed.

Family letter* (p. 169): 1 per student. Remember to sign and date the letter before copying.

Quick Images* (p. 201): 1 transparency of each sheet. Cut out the images on each transparency and store each set in a separate resealable plastic bag or envelope. Label sets as Dot Patterns, 10 Frames, Rectangular Arrays, and Dot Arrays. These images will be used throughout the unit. **Note:** If an overhead projector is unavailable, draw the images on heavy paper to make Quick Image Cards.

Shape Cards (p. 205): 2 sets for use with the class,* plus 1 of each per group of 2–3 students. The Shape Cards will last longer if you duplicate them on heavy paper and laminate them. Sets can be stored in envelopes or resealable plastic bags. (These cards will be used again in Investigation 2.)

Student Sheet 2, Shapes at Home (p. 171): 1 per student (homework)

Continued on next page

For Sessions 2–3

Student Sheets 3–6, Predict and Cover (pp. 172–175): 1 per student, plus a transparency of Student Sheet 3*

Student Sheet 7, Build the Geoblock (p. 176): 1 per pair

Student Sheet 8, Shapes Within Shapes (p. 177): 1 per student (homework)

For Sessions 4–5

Student Sheet 9, Solve Puzzles Recording Sheet (p. 178): 1 per pair, plus a transparency*

Student Sheet 10, Squares (p. 179): 1 per student (homework). Use oaktag or heavy paper, if possible

Student Sheet 11, Composing New Shapes with 2 Triangles (p. 180): 1 per student (homework)

For Sessions 6–8

Student Sheets 3–6, Predict and Cover (pp. 172–175): 1 shape per pair, plus a transparency of Student Sheet 6*

Student Sheet 12, Build a Building (p. 181): 1 per student, plus a transparency.* **Note:** Student Sheet 12 is designed for use with 3/4" interlocking cubes. If you have 2-cm cubes, you will need to enlarge this sheet by about 5% on a copier. Experiment with one enlarged copy first and test it carefully to be sure the outlines match the faces of the cubes students will be using.

Student Sheet 13, Composing New Shapes with 3 or 4 Triangles (p. 182): 1 per student (homework)

■ Prepare a math folder for each student if you did not do so for a previous unit. (Session 1)

■ Separate each tub of Geoblocks into two equal sets. Each set should be enough for 6–8 students. The easiest way to divide the blocks is to find two identical blocks and put one in each set. Store each set in a container about 12" by 15". You will want to keep the Geoblocks separated in this manner for the entire unit. (Sessions 2–3)

■ If you are using computers with this unit, install *Shapes* on each available computer. Read and try the activities presented in the *Shapes* Teacher Tutorial (p. 129). See the **Teacher Note**, Introducing the *Shapes* Software (p. 36), for suggestions on introducing the software to your class. (Sessions 4–5)

■ Prepare sets of 60 interlocking cubes per pair, stored in resealable plastic bags or plastic tubs with lids. (Sessions 6–8)

Shapes Around Us

What Happens

Students look for two-dimensional shapes in the classroom. They describe shapes from a set of Shape Cards and place together those with the same attribute. They sort the cards in different ways and make a poster to show one way of sorting. Their work focuses on:

- identifying shapes in the environment
- describing two-dimensional shapes
- identifying categories for two-dimensional shapes
- sorting two-dimensional shapes in different ways

Start-Up

Today's Number Today's Number is one of the routines that are built into the grade 2 *Investigations* curriculum. Routines provide students regular practice in important mathematical ideas such as number combinations, counting and estimating data, and concepts of time. For Today's Number, which is done daily (or most days), students write number sentences that equal the number of days they have been in school. The complete description of Today's Number (p. 116) offers suggestions for establishing this routine and some variations.

If you are doing the full-year grade 2 *Investigations* curriculum, you will have already started a 200 chart and a counting strip during the unit *Mathematical Thinking at Grade 2*. Write the next number on the 200 chart and add the next number card to the counting strip. As a class, brainstorm ways to express the number.

If you are teaching an *Investigations* unit for the first time, here are a few options for incorporating Today's Number as a routine:

- **Begin with 1** Begin a counting line that does not correspond to the school day number. Each day add a number to the strip and use this number as Today's Number.
- **Use the Calendar Date** If today is the sixteenth day of the month, use 16 as Today's Number.

Once Today's Number has been established, ask students to think about different ways to write the number. Post a piece of chart paper to record their suggestions. You might want to offer ideas to help students get started. If Today's Number is 45, you might suggest 40 + 5 or 20 + 25.

Materials

- Prepared set of Quick Images: Dot Patterns
- Overhead projector (optional)
- Prepared set of Shape Cards (for demonstration)
- Student Sheet 1 (1 per student)
- Shape Cards (1 set per 2–3 students)
- Scissors
- Large paper (1 sheet per group)
- Paste or glue sticks
- Family letter (1 per family)
- Math folder (1 per student)
- Student Sheet 2 (1 per student, homework)

Ask students to think about other ways to make Today's Number. List their suggestions on the chart paper. As students offer suggestions, occasionally ask the group if they agree with the statements. This gives students the opportunity to confirm an idea that they had or to respond to an incorrect suggestion.

As students grow more accustomed to this routine, they will begin to see patterns in the combinations, have favorite kinds of number sentences, or use more complicated types of expressions. Today's Number can be recorded daily on the Weekly Log. (See p. 10.)

Quick Images: A New Routine

Quick Images is a new routine introduced in this unit and is meant to be used in addition to Today's Number and How Many Pockets? Like those routines, Quick Images should be done regularly. However, the intervals at which the three routines are done vary (Today's Number is a daily activity). Consider using Quick Images three times a week at the beginning of math time. See the complete description of Quick Images (p. 125), for suggestions for establishing this routine and some variations.

Use Quick Images: Dot Patterns, 10 Frames, Dot Arrays, and Rectangular Arrays to make four transparencies. (See Other Preparation, p. 3 for further information.) Begin with patterns from Dot Patterns. Describe the activity to students.

I'm going to show you a picture of some dots for 3 seconds. Then I'll cover the picture and ask you to tell how many dots there are.

Note: The following sequence can also be found on p. 125 in greater detail.

- Begin by flashing an image such as five or six dots for 3 seconds. Encourage students to look at the dot arrangement carefully while it is visible.
- With the picture covered (or Quick Image card hidden from view), ask students to tell how many dots there are and to describe what they saw.
- Flash the image again for 3 seconds, then hide it again. This gives students an opportunity to adapt their visual images.
- Ask students to tell how many dots they think there are.
- Display the pattern one more time. Leave it visible and ask students how many dots there are. Encourage them to explain how they reached their answer.

Continue the activity with several other Dot Pattern cards.

Shapes in the Classroom

This activity can be used to introduce the investigation.

In this investigation we are going to be looking at shapes and how we can put them together and take them apart to make other shapes. Sometimes we will be using materials we have used before. Sometimes we will look at shapes in the room or at home.

Ask students to look around and see if they can find any shapes in the classroom such as squares, rectangles, or triangles. Encourage students to look for shapes in other shapes.

I'm looking at the window. When I look at the edge all around the window, I see a big rectangle. When I look at the panes in the window, I see eight squares. Can you see what I see? Who can show the shapes on the window? Can you find shapes other places in the room?

Ask students to share their observations about shapes they see in the classroom—windows, doors, books, the clock, and so on. Continue the discussion for several minutes or for as long as students' observations are sustained.

Sorting Shape Cards

For this activity you will need the prepared set of Shape Cards. Gather students in a circle so they can see the cards.

The cards in this envelope have different shapes. As I come around, reach in, take out one card, and place it in front of you. Look at your shape and think about how you can describe it.

When all students have a card, ask a volunteer to place his or her Shape Card in the center of the circle and tell one thing about the shape. For example, a student with Shape L might say, "My shape has one long side." Students with shapes that match the description place their shapes in the center. For this rule, these might include Shapes D, E, I, J, P, and R. Students with shapes that don't match place their shapes in front of them so they can be seen by everyone.

❖ **Tip for the Linguistically Diverse Classroom** Encourage students to point to each attribute as they describe their shape.

Look at all the shapes in the center. Are they all [triangles]? Look at all the shapes that are not in the center. Are all of them [not triangles]?

Based on the discussion, students may want to make adjustments on how they placed their shapes. When everyone agrees that the shapes are sorted correctly, students retrieve their shapes and place the cards in front of them. Repeat the activity several times. Each time, the student selected must describe one thing about his or her shape that hasn't been mentioned before.

Distribute a set of Shape Cards to each group of two to three students. Students cut out the cards to make their set.

Sort your Shape Cards in at least two different ways. Sort them into two, three, or four groups so that every card has a place. Later you'll make a Shape Card poster to show one way that you've sorted.

Observing the Students Walk around and observe students while they work. Encourage students to place all the cards in the set into categories. If students have difficulty getting started, ask questions such as the following.

- **Which shapes seem to go together?**
- **If you're sorting this way, where does this shape go?**
- **What do you call this group of shapes?**
- **Why did you place this shape in this group instead of that one?**

Categorizing is a complex process. It involves identifying ways that things are the same or different. Identifying a category requires deciding which characteristics you are going to pay attention to and which are irrelevant. For a category to be meaningful, it must include some things and exclude others. As seven- and eight-year-olds classify shapes, they are just learning about these ideas. Although a triangle may seem like a simple shape, it has many different characteristics, any of which a student might focus on: size, the shape of the corners, the number of sides, the proportion of its height to its base (is it *fat* or *skinny*?), and so on. See the **Dialogue Box**, Categorizing Shape Cards (p. 12), for examples of how students in one second grade classroom sorted the cards.

Making Shape Card Posters

When students have sorted the shapes in different ways, distribute a sheet of large paper to each group.

Choose one way that you sorted the Shape Cards. Make a poster by gluing the Shape Cards on the paper in their groups. Be sure to label each group of Shape Cards on your poster.

Students arrange the set of Shape Cards on paper, then glue the cards in place. They write the category name beside each group of cards. Students share their posters.

Introducing Math Folders and Weekly Logs

If you are using the full-year *Investigations* curriculum, students will be familiar with math folders and Weekly Logs. If this curriculum is new to students, tell them about one way they will keep track of their math work.

Mathematicians show how they think about and solve problems by talking about their work, drawing pictures, building models, and explaining their work in writing so that they can share their ideas with other people. Your math folder will be a place to collect the writing and drawing that you do in class.

Distribute math folders to students and have them label them with their names.

Your math folder is a place to keep track of what you do each day in math class. Sometimes there will be more than one activity to choose from, and at other times, like today, everyone in the class will do the same thing. Each day you will record what you did on this Weekly Log.

Distribute Student Sheet 1, Weekly Log, and ask students to write in their names at the top of the page. Point out that there are spaces for each day of the week and ask them to write today's date on the line after the appropriate day. If you are doing the activity Today's Number, students can write the daily number in the box beside the date.

Ask students for suggestions about what to call today's activities. Titles for choices and whole-class activities should be short to encourage all students to record what they do each day. List their ideas on the board and have students choose one title to write in the space below the date.

❖ **Tip for the Linguistically Diverse Classroom** Encourage students who are not writing comfortably in English to use drawings to record in their Weekly Log. If students demonstrate some proficiency in writing, suggest that they record a few words with their drawings. Students can also record a sample problem representative of each day's work.

Weekly Logs can be stapled to the front of the folders (each new week on the top so students can view prior logs by lifting up the sheets).

During the unit (or throughout the year), you might use the math folders and Weekly Logs in a number of ways:

■ to keep track of what kinds of activities students choose to do and how frequently they choose them

■ to review with students, individually or as a group, the work they've accomplished

■ to share student work with families, by sending folders home periodically for students to share or during student/family/teacher conferences

Session 1 Follow-Up

Shapes at Home Students look around at home to find at least five examples of different shapes. Students write about or draw a picture of what they find on Student Sheet 2, Shapes at Home.

Send home the family letter or the *Investigations* at Home booklet.

Homework

D I A L O G U E ☐ B O X

Categorizing Shape Cards

The students are working in pairs to sort their Shape Cards into categories. In a discussion during the activity Sorting Shape Cards (p. 7), the teacher checks in with each pair and talks with them about the characteristics of each of their groups.

Lila and Rosie group one or two shapes at a time, define a category that describes them, then move on to another small group of shapes, until all the shapes are categorized. So far they have five categories: squares, triangles, diamonds, big, and small.

Could any of the diamonds go in big or in small?

Lila: We want them to be diamonds.

How would I know which group to put this one in [*points to Shape M, the small square rotated 90 degrees*]?

Rosie: Well, it is small, but it's a diamond, too.

Lila: Small can be just anything, but diamonds have to go in diamonds.

Rosie: We can put arrows between small and diamonds so people would know it can go in both.

The teacher moves to Imani and Ping, who have used two characteristics for each of their categories. They have labeled their groups LONG AND POINTY PILE, BIG AND FAT PILE, LONG AND SKINNY PILE, and DIFFERENT PILE. The DIFFERENT PILE seems to have everything they couldn't fit into another category.

Another pair, Graham and Carla, have two categories so far, POINTY and BIG.

How would I know if something is POINTY?

Carla: Like this. [*She holds up a triangle and points to a corner.*]

Could a rectangle go in this category?

Carla: No. Points aren't like that. They don't go straight. They have to go up.

Graham: [*holds up two rectangles*] What do you call these?

Which ones?

Graham: The kind with the short and long sides.

Carla: Rectangles. If we said three sides and four sides, we'd be all done. [*She shows how she could sort the shapes into two big piles of three-sided and four-sided shapes.*]

Graham: [*picks up Shape I, the rectangle with the corner cut out, which Carla has placed in the four-sided pile*] I don't think this one goes.

Carla: Well, it can't go in the threes.

How many sides does it have?

Carla: Four, except it has a bite out of it. It can go with these [*indicates the four-sided group*]. It's close enough.

What do you think, Graham?

Graham: It doesn't go there, either. Look. It has one, two, three, four, five, six sides [*he traces each side with his finger*].

Carla: But it's really just like these. It's just like you ripped out the corner.

Students at this age may often choose categories that are not mutually exclusive, so that some of the shapes they are using will fit into more than one category. While an adult might sort either into big and small or into triangle and quadrilateral, choosing to focus on either size or shape, students often mix categories. For example, Lila and Rosie use big, small, triangle, and square in the same sort. However, as students make decisions and describe their categories, they begin to

Continued on next page

continued

make important distinctions, as Carla does when she describes the difference between right angles that "go straight" and acute angles that "go up."

Some students will focus more on their impression of the whole shape, as Imani, Ping and Carla do. For Carla, the rectangle with the missing corner still looks basically like a rectangle—and, from a certain point of view, she is correct about its overall similarity to a rectangle. Graham can focus on the number of sides as a separate and important characteristic of this rectangle like shape. Throughout these activities, what is important is that students are challenged to articulate as clearly as they can what criteria they are using to make their category decisions.

Seeing Shapes Within Shapes

Materials

- Pattern blocks (1 tub per 6–8 students)
- Chart paper or poster board
- Student Sheets 3–6 (1 per student)
- Transparency of Student Sheet 3 (optional)
- Overhead projector (optional)
- Prepared set of Geoblocks (1 per group)
- Student Sheet 7 (1 per pair)
- Student Sheet 8 (1 per student, homework)

What Happens

Students find all the ways to use other blocks to cover each of the pattern blocks. These are recorded on a class chart. They also put Geoblocks together to form other blocks. They work on two Choice Time activities: Predict and Cover (using pattern blocks) and Build the Geoblock. Their work focuses on:

- fitting shapes together to cover a region
- putting parts together to form a whole
- combining three-dimensional shapes to make a three-dimensional whole

Start-Up

Today's Number Suggest that students use combinations of 10 in their number sentences. For example, if Today's Number is 80, a possible number sentence includes (6 + 4) + (5 + 5) + (7 + 3) + (5 + 5) + (8 + 2) + (1 + 9) + (3 + 7) + (4 + 6). Add a card to the class counting strip and fill in the next number on the blank 200 chart. For complete details on this routine, see p. 116.

Shapes at Home Students share examples or pictures of some of the shapes they found at home.

Activity

Covering Pattern Blocks

Distribute pattern blocks to each group of students. If students have not used them recently, provide free exploration time before introducing a task. See the **Teacher Note**, Identifying Pattern Blocks and Terminology (p. 22), for a description of the blocks and for ways students might identify them.

Make some of the pattern block shapes by putting other blocks together. Look at the red trapezoid, for example. You can make that shape using some of the other blocks. Try it by building on top of the trapezoid or right beside it. Who found a way to do it?

Accept one example, such as placing three green triangles on top of a red trapezoid.

See which pattern block shapes you can build with other blocks. **Find all the ways to make each shape.** Every time you find a way, leave it on the table [or desk, or floor]. For example, can you find ways to put other pattern blocks together to make a hexagon the same size as the yellow one? Can you find ways to put other blocks together to make a square the same size as the orange one? Can you do it for the trapezoid? the triangle? the two diamonds?

Observing the Students Walk around and observe students as they work. If they have difficulty understanding the task, demonstrate what they are to do, using the example suggested for the red trapezoid.

■ Do students try to find several ways to cover a pattern block, or do they try covering another block after finding just one way?

■ Do students share and compare their ways with others in their group?

When most of the students are finished, discuss the activity with the class. Students should look at and use their arrangements as they share results. Record their results on a class chart that lists (or shows in a drawing) all six pattern blocks. (Use chart paper or poster board.)

Which pattern block do you want to start with? [*Students choose the hexagon.*] **Who has found a way to exactly cover the yellow hexagon?** [*Record results on the chart using words, symbols, or drawings.*] **Who has found a different way?**

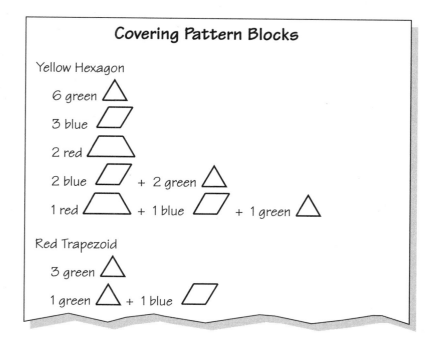

Continue until students have described all the ways they have found, recording after each response, and then go on to a different block. Leave the chart posted so students can add any further arrangements they find.

Predict and Cover

In this activity, students will use the relationships they've discovered among the pattern blocks to make predictions about how many blocks it takes to cover shapes. Distribute pattern blocks and Student Sheet 3, Predict and Cover (Shapes A and B), to students. Place a transparency of the student sheet on the overhead.

Look at Shape A. How many blue rhombuses do you think it will take to cover it? Try to make a prediction without using the blocks.

Ask several students to share their predictions. Then students check their predictions by covering Shape A with blue rhombuses.

You've found that it takes five blue rhombuses to cover Shape A. How many green triangles do you predict it will take to cover it?

Some students may be able to use the relationship between rhombuses and triangles to figure out how many triangles will cover the shape. They will "see" two triangles in each of the places in Shape A where a rhombus can fit, and will then count the triangles that they imagine (saying, for example, 1, 2; 3, 4; 5, 6; and so on). A few students may analyze the numerical relationships: If it takes 5 rhombuses to cover Shape A and 2 triangles to cover a rhombus, it must therefore take 10 triangles to cover Shape A.

Other students will be unable to use the relationship between triangles and rhombuses to predict how many triangles cover Shape A. They will figure out the number of triangles by trying to visualize where each triangle fits. These students will have more experiences thinking about how relationships between blocks can be used during the Choice Time activities.

Look at Shape B. Is it easier to predict how many red trapezoids or how many triangles will cover the shape? Why? How many trapezoids does it take to cover the shape?

Students record their predictions on the student sheet and then cover the shape with trapezoids and record their counts. Next, ask students to predict how many triangles will cover the shape. As students share their predictions, notice whether they mention the relationship between trapezoids and triangles. Students can go on to cover the shape with triangles, then count.

Students will be continuing this activity during Choice Time using Student Sheets 4–6.

Build the Geoblock

Geoblocks were introduced in the unit *Mathematical Thinking at Grade 2*. If this material is new to students, provide time for free exploration before doing specific activities. See the **Teacher Note**, Geoblocks (p. 23), for a description of the material.

This activity can be briefly demonstrated and explained to the class. Students will continue this activity during Choice Time. You may need to spend some time with small groups of students to help them get started.

Introduce this activity by telling students that they will be putting Geoblocks together to build other blocks. Hold up (or place on a table where all can see) the rectangular prism that measures 8 cm by 4 cm by 4 cm.

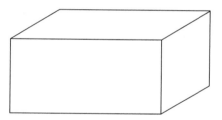

If I could glue other Geoblocks together, how could I build a block that is the same size and shape as this one? One way would be to glue these two blocks together. [*Hold up two 4-cm cubes to show that they could be glued together to make the same size and shape block.*]

Display Student Sheet 7, Build the Geoblock, and demonstrate how to find the Geoblocks that are illustrated—the front faces of the blocks fit the drawings. Explain that for each of the three blocks on the sheet, students are to find the Geoblock that fits and then see how many different ways they can put other Geoblocks together to make a block the same size and shape. They should keep all their buildings until you have an opportunity to see what they have done. Tell them to find more than two or three ways for each block—there are many ways to put other Geoblocks together to make the shapes.

If you introduce this activity to small groups, follow the above procedures except ask students to find one or two ways to build the block (rather than show them the two 4-cm cubes). Each group of students will need one set of Geoblocks—a half set of the tub of blocks. Groups take their set of blocks to different locations. Remind students to keep their block constructions together until you have checked with them.

Introducing Choice Time

Choice Time is a format that recurs throughout the grade 2 *Investigations* curriculum. See the **Teacher Note**, About Choice Time (p. 24), for information about how to set up Choice Time, including how students might use their Weekly Log to keep track of their work.

Explain how students are to work during Choice Time. Each day, they select one or two of the activities they want to participate in. Students can select the same activity more than once, but they should not do the same activity each day. Decide whether you want students to do every activity or just some of them.

List these choices on the chalkboard.

> 1. Predict and Cover
>
> 2. Build the Geoblock

If Choice Time is new for your class, you may need to help students plan their activities. Assure them that they will have the opportunity to try each choice. Support students in making these decisions and plans for themselves rather then organizing them into groups and circulating the groups. Making choices, planning time, and taking responsibility for their own learning are important aspects of a student's school experience.

Choice 1: Predict and Cover

Materials: Pattern blocks; Student Sheets 4–6, Predict and Cover

Students continue the work begun during the activity Predict and Cover. For each outline of Shapes C, D, E, F, G, and H (Student Sheets 4–6), students predict the number of blocks that will cover the shape, then check their prediction. Ask students to do the shapes in order and finish each shape outline, covering it with the blocks indicated, before going on to the next shape.

On Student Sheets 4 and 5 (Shapes C, D, E, and F), the particular blocks to use are given. For Student Sheet 6 (Shapes G and H), students look at each outline and decide which block is the easiest to predict, then draw a picture or write the name of the block on the block line for the first prediction. For example, students may say that for Shape G, it is easiest to predict the number of rhombuses. After predicting and covering with one pattern block shape, students predict and cover an outline using a second pattern block shape and then, perhaps, a third block shape.

Students should finish this activity during Sessions 2 through 5 so you can discuss their results at the end of Session 5. Students do not need to cover every shape, but encourage them to do at least four.

Choice 2: Build the Geoblock

Materials: Geoblocks; Student Sheet 7, Build the Geoblock

For each of the three blocks pictured on the student sheet, partners find the Geoblock that matches and then see how many different ways they can put other Geoblocks together to make a block the same size and shape. Ask them to keep all their buildings until you see what they have done. Encourage students to find more than two or three ways for each block—there are many ways to put blocks together to make each shape.

To ensure that blocks are accessible, no more than four students should share one (half) set of blocks. Students work with partners to do one of the three tasks on the student sheet. When pairs have completed several (or all) of the buildings for a task, they call the teacher over to check their work.

If students finish building the blocks on the student sheet, they can find different ways to build other Geoblocks.

Observing the Students

As students work on Choice Time activities, you will have the opportunity to observe and listen to them. The **Teacher Note,** Keeping Track of Students' Work (p. 26), suggests ways to record and use your observations.

Do students try each choice, or do they stay with a familiar one? If after a short time students say they're finished, ask them to tell you what they have done and encourage them to investigate further. Notice whether students work alone or with partners. Do they share what they have done with others and observe what others are doing? Do they talk to themselves or others about what they are doing?

The following pages offer some specific suggestions of what you might observe as students work on Choice Time activities.

Predict and Cover

- Do students start with any block, or do they try the blocks in the specified order?

- On Student Sheet 6, do they choose a block they think will be easy to predict, or do they start with the smaller blocks?

- After covering a shape with one block, do students use this information to help them to predict the count for a different block? See the **Dialogue Box,** Predict and Cover (p. 29), for how a teacher encourages students to use known information for predicting. Remind students to record their predictions before covering a shape.

Note: Students may be so concerned that their written answers be correct that they will change their predictions to match actual counts. Explain to students that predictions do not need to be correct. Talk about what a prediction is, helping students come to see it as a thoughtful guess. Reassure them that predictions are often not the same as the actual count.

As you interact with students during Choice Time, encourage them to use relationships among the pattern blocks to make better predictions. Keep in mind, however, that use of such relationships will vary greatly among students and among shapes. It will probably be more difficult for students to use relationships between blocks as they work on Shapes F, G, and H. For more information, see the **Teacher Note,** Seeing Relationships Between Shapes (p. 27).

Students who are having difficulty making predictions can be helped in several ways.

- When students are predicting how many of a small block will cover a shape, leave their covering of the shape with a larger block visible. Ask, "Can knowing about how many of these covered the shape help you predict?"

- If students get mixed up counting the actual number of small blocks that cover a shape, have them count the blocks as they remove them one at a time.

- After students have covered about half of a shape with small blocks, ask how many they predict will cover the whole shape. Because they have already covered half the shape, students are often able to make better predictions.

- To predict, have students try to draw where they think the shapes will fit.

After giving a student a hint on one problem, try not to give one on the next. Also keep in mind that the strategy that students use often depends on the problem.

Build the Geoblock

Since students will not be recording their work, periodically check with students who are working on this choice. Encourage pairs to find many ways to build a block before going on to another block. You may want to have all students who are using a set of Geoblocks work together to find ways to build the same block. Check their work when they have completed several (or all) of the buildings for a block. Ask them to tell you how they made each building, including the number of blocks they used and an informal description of the blocks.

Observe how students put the blocks together. Although this is an exploratory activity, you can learn a great deal about how students perceive three-dimensional shapes by observing how they attempt to build them.

■ Do students randomly choose smaller blocks and try to build a larger one?

■ Do they seem to "see" blocks that are half of a block and put two of them together?

■ Do they use smaller cubes to build the larger cube?

Near the End of the Session Five or 10 minutes before the end of each Choice Time session, have students stop working, put away the materials they have been working with, and clean up their work areas. Ask everyone to double-check the floor for pencils, stray blocks, and other materials.

When cleanup is complete, students record their daily work in their logs. Suggest that they use the list of activities that you posted as one reference for writing what they did during Choice Time.

Whenever possible, either at the beginning or end of Choice Time, have students share some of the work they have been doing. This often sparks interest in an activity. Some days you might ask two or three students to share with the class the work they have been doing. On other days you might ask a question that came up during Choice Time so that others can respond to it. Sometimes you might want students to explain how they thought about or solved a particular problem.

Sessions 2 and 3 Follow-Up

Shapes Within Shapes   Students look for shapes within shapes at home. For example, they might find a window that is a rectangle and is made up of four rectangles. They write or draw at least three shapes within shapes on Student Sheet 8, Shapes Within Shapes.

Homework

Identifying Pattern Blocks and Terminology

The pattern block set is made up of six shapes: a hexagon, a trapezoid, a square, a triangle, and two parallelograms (rhombuses). In most sets each shape comes in one color: the hexagons are yellow; the trapezoids are red; the squares are orange; the triangles are green; the narrow rhombuses are tan; and the wide rhombuses are blue.

Since all pattern blocks of the same color are the same shape, it is very natural for students to identify them by color. This is fine and should not be discouraged. At the same time, students should become familiar with and use correct geometric terms for different shapes. Second grade students will easily identify the green block as a triangle and the orange block as a square. The terms *trapezoid* and *hexagon* may be new to many students. They will learn to use these words readily if you use them naturally and help them remember the words when they forget. To help students learn the names of the shapes, you may want to use the color along with the shape name for a while, such as yellow hexagons and red trapezoids.

How to identify the blue and tan blocks becomes a little more problematic since there is no unique term that applies to these blocks. Young children typically refer to them as diamonds. Along with the orange squares, they are rhombuses, and they are also parallelograms. This is because many geometric terms for shapes are part of a hierarchical classification system. For example, all squares are rectangles, but not all rectangles

are squares. The defining characteristic of four-sided shapes are:

- Square: 4 equal sides and 4 right angles
- Rhombus: 4 equal sides
- Rectangle: 4 right angles
- Parallelogram: 2 pairs of parallel sides
- Trapezoid: 4 sides and only two sides parallel
- Quadrilateral: 4 sides

The blue and tan blocks are identified as rhombuses. (The tan block is identified as the thin rhombus.) When you talk about these shapes, frequently use the term *parallelogram* so students become familiar with both terms. Identifying the blue block as a rhombus and the tan block as a parallelogram may lead to misconceptions about the differences between the two blocks. Students will still need to distinguish among the different blocks that are rhombuses, parallelograms, or diamonds.

Model the correct use of geometric terms but do not insist that students use them. As long as they are communicating effectively, let them use the language they are comfortable with, while you continue to model the geometric language. You should be cautious while using the pattern blocks that you don't inadvertently lead students to misconceptions. For example, if a student says, "None of the blocks are rectangles," you might respond, "None of the blocks look like many of the rectangles you have seen. However, a square is a special kind of rectangle."

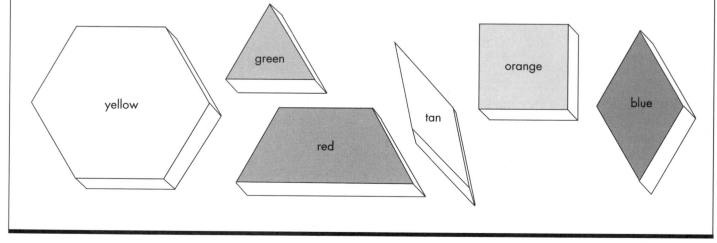

Geoblocks

Geoblocks are three-dimensional wooden blocks. While the blocks look similar to kindergarten blocks, they are smaller and are related by volume. There are also several different sizes for many of the shapes. For example, the set includes 1-cm, 2-cm, and 3-cm cubes.

Most second graders love to build with these blocks—they build towers, towns, roads, ramps, bridges, and many other things. As with other manipulative materials, such as pattern blocks, most students will need time to explore the Geoblocks before they are ready to use them in more specified ways. During this informal building time, they intuitively learn many of the characteristics of the blocks. They may discover, for example, that they need to substitute two smaller blocks to use in their road when the larger blocks run out.

In real-life situations, we frequently use two-dimensional drawings to help us picture and represent three-dimensional things. For example, a blueprint provides instructions for building a house, a pattern provides instructions for cutting out and sewing a shirt, and a diagram provides instructions for assembling a bike. One reason Geoblocks are included in the second grade materials is to provide students with the opportunity to work with three-dimensional materials and see the relationship between three- and two-dimensional shapes.

In the introductory unit of the grade 2 *Investigations* curriculum, students examined Geoblocks and described their attributes, sorted the set, and found Geoblocks that match their two-dimensional faces. In addition, students counted the number of different Geoblocks. If students have not done these activities, you may wish to have them spend some time doing them so they become more familiar with the various blocks in the set.

Choice Time is an opportunity for students to work on a variety of activities that focus on similar mathematical content. Choice Time sessions are found in most units of the grade 2 *Investigations* curriculum. These generally alternate with whole-class activities where students work individually or in pairs on one or two problems. Each format offers somewhat different classroom experiences. Both are important for students to be engaged in.

In Choice Time the activities are not sequential. As students move among them, they continually revisit some of the important concepts and ideas they are learning. Many Choice Time activities are designed with the intent that students will work on them more than once. By playing a game a second or third time or solving similar problems, students refine strategies, see a variety of approaches, and bring new knowledge to familiar experiences.

You may want to limit the number of students who work on a Choice Time activity at one time. Often when a new choice is introduced, many students want to do it first. Assure them that they will be able to try each choice. In many cases, the quantity of materials available will limit the number of students that can do an activity at any one time. Even if this is not the case, set guidelines about the number of students who work on each choice. This gives students the opportunity to work in smaller groups and make decisions about what they want and need to do. It also provides a chance to return and do some choices more than once.

Initially you may need to help students plan what they do. Rather than organizing them into groups and moving the groups to a new activity periodically, support students in making decisions about the choices they do. Making choices, planning their time, and taking responsibility for their own learning are important aspects of a student's school experience. If some students return to the same activity over and over again without trying others, suggest that they make a different first choice and then choose the favorite activity as a second choice.

How to Set Up Choices

Some teachers prefer to have choices set up at centers or stations around the room. At each center, students find the materials needed to complete the activity. Other teachers prefer to have materials stored in a central location and have students bring materials to their desks or tables. In either case, materials should be readily accessible to students, and students should be expected to take responsibility for cleaning up and returning materials to their appropriate storage locations. Giving students a "5 minutes until cleanup" warning before the end of an activity session allows them to finish what they are working on and prepare for the upcoming transition.

You may find that you need to experiment with a few different structures before finding a setup that works best.

The Role of the Student

Establish clear guidelines when you introduce Choice Time activities. Discuss students' responsibilities:

- Try every choice at least once.

- Work with a partner or alone. (Some activities require that students work in pairs, while others can be done either alone or with partners.)

- Keep track, on paper, of the choices you have worked on.

- Keep all your work in your math folder.

- Ask questions of other students when you don't understand or feel stuck. (Some teachers establish the rule, "Ask two other students before me," requiring students to check with two peers before coming to the teacher for help.)

Continued on next page

Students can use their Weekly Logs to keep track of their work. As students finish a choice, they write it on their log and place any work they have done in their folder. Some teachers list the choices for sessions on a chart, the board, or the overhead projector to help students keep track of what they need to do.

In any classroom there is a range of how much work students complete. Some choices include extensions and additional problems for students to do when they have completed their required work. Encourage students to return to choices they have done before, do another problem or two from the choice, or play a game again.

At the end of a Choice Time session, spend a few minutes discussing with students what went smoothly, what sorts of issues arose and how they were resolved, and what students enjoyed or found difficult. Encourage students to be involved in the process of finding solutions to problems that come up in the classroom. In doing so, they take some responsibility for their own behavior and become involved with establishing classroom policies. You may also want to make the choices available at other times during the day.

The Role of the Teacher

Choice Time provides you with the opportunity to observe and listen to students while they work. At times, you may want to meet with individual students, pairs, or small groups who need help, or whom you haven't had a chance to observe before, or to do individual assessments. Recording your observations of students will help you keep track of how they are interacting with materials and solving problems. The **Teacher Note,** Keeping Track of Students' Work (p. 26), offers suggestions for recording and using your observations.

During the initial weeks of Choice Time, most of your time will probably be spent circulating around the classroom helping students get settled into activities, and monitoring the overall management of the classroom. Once routines are familiar and well established, students will become more independent and responsible for their work. This will allow you to spend more concentrated periods of time observing the class as a whole or working with individuals and small groups.

Throughout the *Investigations* curriculum, there are numerous opportunities to observe students as they work. Teacher observations are an important part of ongoing assessment. Individual observations are snapshots of a student's experience with a single activity. When considered over time, a set of observations can provide an informative and detailed picture. These observations can be useful in documenting and assessing a student's growth. They offer important information when preparing for family conferences or writing student reports.

Your observations of students will vary throughout the year. At times you may be interested in particular strategies that students are developing to solve problems. Sometimes you might want to observe how students use or do not use materials to help them solve problems. Or you may be interested in noting the strategy that a student uses when playing a game during Choice Time. Class discussions also provide many opportunities to take note of students' ideas and thinking.

Keeping observation notes on a class of 28 students can become overwhelming and time-consuming. You will probably find it necessary to develop a system to record and keep track of your observations of students. A few ideas and suggestions are offered here, but you will want to find a system that works for you.

A class list of names is convenient for jotting down observations of students. Since space is somewhat limited, it is not possible to write lengthy notes. However, when kept over time, these short observations provide important information.

Stick-on address labels can be kept on clipboards around the room. Notes can be taken on individual students and then these labels can be peeled off and stuck in a file that you set up for each student.

Alternatively, you might find that jotting down brief notes at the end of each week works well for you. Some teachers find that this is a useful way of reflecting on the class as a whole, on the curriculum, and on individual students. Planning for the next weeks' activities often develops from these weekly reflections.

In addition to your own notes on students, each student will be keeping a folder of work. This work and the daily entries on the Weekly Logs can document a student's experience. Together they can help you keep track of the students in your classroom, assess their growth over time, and communicate this information to others. At the end of each unit there is a list of things you might keep in students' folders.

Seeing Relationships Between Shapes

The goal of many of the activities in this investigation is to help students mentally compose and decompose shapes. Visualizing how a larger unit can be composed of smaller ones and how shapes can be combined to form larger units is the basis of understanding many geometric ideas, from relationships among polygons (can all rectangles be cut into two congruent triangles?) to area and volume. When students can put shapes together to form a new shape, and then see that new shape as a new unit with which they can cover or measure, they have created a *composite unit*. For example, suppose a student is asked to predict how many rhombuses it takes to cover the shape below.

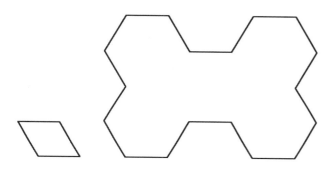

If a student forms a hexagon from three rhombuses and then thinks of covering the shape with copies of this set-of-rhombuses-as-hexagon, then this set of rhombuses is being used as a composite unit.

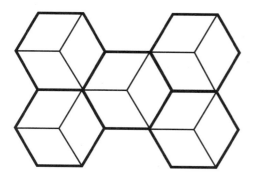

Use of these mental processes enables students to see that they can use relationships among blocks to find answers to other shape problems. For example, in the previous problem, using the threesome of rhombuses leads some students to use skip counting to determine the number of rhombuses that cover the shape—3, 6, 9, 12, 15.

The activities in this investigation will provide you with many opportunities to understand how students are thinking. We have observed several levels of sophistication in students' solutions to the Predict and Cover problems.

Doesn't use previous information The student doesn't use relationships between blocks to help solve the problems. For example, Carla covers Shape D with five hexagons, but she doesn't use that information to figure out the number of blue rhombuses that cover Shape D. She draws with the eraser side of her pencil while counting 1, 2, 3, 4, 5, 6, 7, 8, 9, 10, 11, 12, 13.

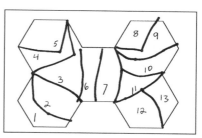

Partially uses previous information The student uses the previously obtained number of smaller blocks it takes to cover a larger block but fails to maintain an image of how the larger block covers the shape. (See Bjorn's response.) Or the student uses information about the number of larger blocks it takes to cover the shape but fails to maintain an image or count of how many smaller blocks cover the larger block. (See Ayaz's response.)

Bjorn: [*makes a rhombus out of 2 triangles*] It takes 2 of these to make 1 of these.

Continued on next page

[*He moves his finger inside the shape, skip counting by 2's as he points to the positions.*]

Bjorn: 2, 4, 6, 8, 10. Maybe 10.

Bjorn was trying to treat a rhombus as a composite unit of 2 triangles. He was unsuccessful because he could not hold in his mind an image of where the 4 embedded rhombuses could be placed.

Ayaz: [*draws in the hexagons*] Then it would make maybe 4. [*He counts each hexagon as two rhombuses.*] 2, 4, 6, 8.

Ayaz saw that 4 hexagons covered the shape. However, he counted each hexagon as 2 rhombuses instead of 3.

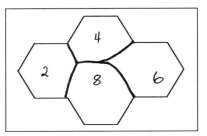

Uses previous information to reach a complete solution The student correctly uses the number of smaller blocks that fit in the larger block as a composite unit to solve the problem.

Bjorn enumerates blue rhombuses in the shape below.

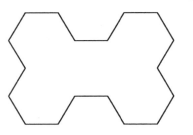

Bjorn: Since it took 3 [rhombuses] to cover 1 of these [hexagon], that would be 3. [*Then he points to the 2 most leftward hexagons in the shape.*] 6. [*He points to the middle.*] This one makes 9 [*points to the upper right*], 12 [*points to the lower right*], 15.

Other students are discussing Shape E. They have established that it takes 4 hexagons to cover it.

Now that you know that it has 4 hexagons, how many trapezoids would you need?

Franco: 8, because I know 2 trapezoids make a hexagon, so I doubled the number.

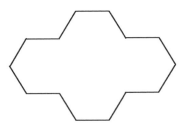

How about green triangles?

Karina: 24.

Why?

Karina: Because 6 green triangles make a yellow hexagon. 6 and 6 is 12, and 12 and 12 is 24.

Laura: I added up eight 3's because there are 8 trapezoids and 3 triangles make a trapezoid.

How about blue rhombuses?

Jeffrey: 3 blues make a hexagon, and 3 times 4 is 12.

Predict and Cover

In this discussion during Choice Time (p. 20), several students have just started Predict and Cover. They are working on covering Shape E with pattern blocks.

Laura and Trini have just found that 4 hexagons will cover the shape and are now predicting how many trapezoids will cover it.

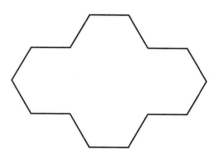

Laura: I think it might be 9 or 10 trapezoids because there were only 4 hexagons and these are smaller so it would take more.

Trini: I think it would be 8 trapezoids because it takes 2 trapezoids to make a hexagon.

Which do you think it will be: 8, 9, or 10 trapezoids?

Laura: [*points and counts*] 1, 2, 3, 4, 5, 6, 7, 8, 9. I think it's 9.

What do you think, Trini?

Trini: I still think it would be 8. There's 4 hexagons, and each one has 2 trapezoids.

Trini, show Laura where you think the trapezoids fit.

[*Trini points and counts as follows.*]

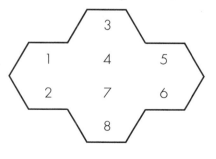

Laura: But when I counted, I got 9.

Suppose we put the four hexagons back on the shape. [*The girls do so.*] **How do the trapezoids fit on the hexagons?**

[*Trini places two trapezoids on top of one of the hexagons.*]

Laura: Oh, I see now. There's going to be 2 more here, and here, and here [*pointing to the three uncovered hexagons*]. I agree with Trini, it's 8.

[*The girls then check with trapezoids.*]

Note how the teacher tried to help Laura see what Trini was able to visualize.

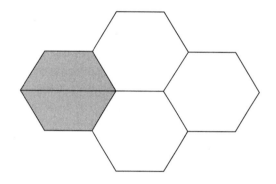

Shape Puzzles

Materials

- Prepared set of Quick Images: Dot Patterns
- Overhead projector (optional)
- Student Sheet 9 (1 per pair)
- Transparency of Student Sheet 9 (optional)
- Computers with *Shapes* installed (optional)
- Large-screen monitor (optional)
- Choice Time materials from previous sessions
- Student Sheet 10 (1 per student, homework)
- Student Sheet 11 (1 per student, homework)

What Happens

Students are introduced to a new Choice Time activity, Solve Puzzles (using the *Shapes* software), and continue to work on the activities introduced previously. In the last half of Session 5, students share their results for Predict and Cover and look for patterns that help them make predictions. Their work focuses on:

- identifying and describing numerical relationships found among the pattern block shapes

Start-Up

Quick Images Flash images of arrangements of dots on the overhead. Use some of the same arrangements you used when you introduced Quick Images and a few new arrangements. Ask students to share why the arrangements of the dots helped them figure out how many were shown. For complete details on this routine, see p. 125.

Today's Number Sometime during the school day, students brainstorm ways to express Today's Number. Add a card to the class counting strip and fill in the next number on the blank 200 chart. See p. 116 for complete details on this routine.

Activity

On-Computer Activity: Introducing Solve Puzzles

The use of computer activities is optional in this unit. However, if you have computers available, it is recommended that you use the software. The computer work is integrated into the units and enriches the work students do. Many of the computer activities cannot be done with manipulative materials.

If you are doing the full-year *Investigations* curriculum, students may be familiar with the *Shapes* software introduced in *Mathematical Thinking at Grade 2*. If students have not used the *Shapes* software, it is important that they become familiar with it before doing activities in this unit. Students need to explore the *Shapes* software and know how to use many of its tools. See the Teacher Tutorial (p. 129) and the **Teacher Note**, Introducing the *Shapes* Software (p. 36), for suggestions on introducing it to your class.

Students may already be familiar with the activity Solve Puzzles if they have used it in the introductory unit and at other times. If not, introduce the activity. In Solve Puzzles, which is similar to Predict and Cover, students fill in outlines on the screen.

The computer activity may be introduced to small groups of students or to the entire class if you have a large-screen display. Gather students around the screen and open the *Shapes* software.

Choose the Solve Puzzles activity. Read the directions, then click on **[OK]** or press **<return>**. Outline Number 1 appears.

Students' task is to fill each outline with shapes. They do not have to use any specific shape, but they should record the shapes and the number of shapes that they used.

Choose **Number 1** from the **Number** menu. Tell students that they need to find the fewest number of blocks it takes to fill the shape and then tell what they are and how many of each block. Show students how to complete the recording sheet for puzzle number 1 using a transparency of Student Sheet 9, Solve Puzzles Recording Sheet. Complete the columns for Shape Number, Fewest Number of Blocks to Fill, and What are they and how many of each block?

Next choose **Number 5** from the **Number** menu. Tell students that puzzles 5 to 10 have two challenges: to fill in each outline with the *fewest* blocks possible, and to *predict* how many *green triangles* to use. Demonstrate how to complete both challenges.

Check the prediction illustrating an easy way to do this task. Use the Glue tool to glue two triangles (one turned or flipped) into a rhombus. Use the Duplicate tool to copy this two-in-one shape repeatedly to fill the outline.

Note: Another way for students to check their predictions about the number of green triangles is to use the Hammer tool until all shapes are hammered to triangles. This might be particularly helpful for those having difficulty visualizing how blocks can be decomposed. In the *Shapes* software for this unit, the hammer not only breaks apart glued groups but also will break single shapes (such as a blue rhombus) into smaller shapes (2 green triangles) whenever possible.

Using the transparency, show students how to complete the recording sheet for puzzle number 5. For Shapes 5–10, students will predict how many green triangles will fill each shape, then check their prediction. Students can continue this activity with partners during Choice Time.

Choice Time

Add Solve Puzzles to the list of Choice Time activities.

---

1. Predict and Cover

2. Build the Geoblock

3. Solve Puzzles (computer)

Your computer setup may require some accommodations. See the **Teacher Note**, Managing the Computer Activities (p. 37), for suggestions on how to structure the computer choice depending on the number of computers you have available.

For a review of the descriptions of Predict and Cover and Build the Geoblock and what to look for as you observe students working on these activities, see p. 18.

Choice 3: Solve Puzzles

Materials: Computers with *Shapes* installed; Student Sheet 9, Solve Puzzles Recording Sheet

Students use the Solve Puzzles activity and fill shapes 1–4 with the fewest blocks possible. They record the number and blocks used on Student Sheet 9. They do the same for shapes 5–10; then they predict and record the number of triangles needed to fill the shape, check their prediction, and record the result. Students do not need to do the shapes in any order, nor do they need to do all the different shapes.

Observing the Students

Students should complete Predict and Cover during the next two sessions. Plan on having a discussion toward the end of Session 5, at which time students can share their results.

Solve Puzzles

As students work on the computer activity, talk with them about the way they are moving the blocks (sliding the blocks and using the Turn and Flip tools). This will help them become more aware of these geometric motions. Just as important, it will help them become familiar with seeing shapes in different orientations and realizing that changing the orientation does not affect the type of shape or its attributes.

- Ask students how they could prove that they have filled each shape with the fewest blocks possible. Are there different ways to do so?
- After filling a shape, what strategies do students use to predict the number of green triangles? For example, do they point to the computer to visualize and count the shapes? Do they figure equivalent amounts for each shape used, such as writing a 6 for each hexagon, and then adding the numbers?

If you wish later to discuss the different ways students filled the shapes, have them save their solutions on disk. See the Teacher Tutorial (p. 165) for information on how to save work on a different disk.

Teacher Checkpoint

Predict and Cover

Teacher Checkpoints are places for you to stop and observe student work. For more information on these features, see About Assessment (p. I-10). However, keep in mind that throughout this entire unit, you will be assessing students' understanding of mathematical ideas.

During the last half of Session 5, when all students have completed Predict and Cover, gather the class together to discuss their results. Students should have pattern blocks and Student Sheets 4–6, Predict and Cover, which they should have completed. (All students may not have covered all shapes.) Before discussing results, review what students were to do on the task.

While you may want to share students' results for all six shapes, choose two or three shapes to discuss in more depth. When discussing a shape, have students fold their student sheets so that only the shape outline shows, with their predictions and results underneath. This procedure will allow students to make visual images of the shapes and blocks for a second time, rather than just referring to their recorded answers. Seeing composite shapes in other shapes comes with experience and familiarity with both the activity and the material.

The following notes may be helpful in guiding your discussion of the different shapes.

Shape C: Some students may spontaneously volunteer that they can see hexagons in the shape. If so, you might ask, "Does it look like we can cover this shape using *only* yellow hexagons?" Encourage students to share their thinking about this question. Some students may see that two hexagons fit in the middle of the shape but that the ends of the shape will not be covered. For other students, this may be more difficult to see.

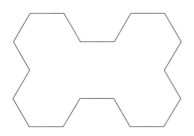

Shape D: Without looking at their results, ask students which pattern block is the easiest to see in the shape and how many of that block will cover it. Some students may not initially see the fifth hexagon in the center of the shape, but it will "pop out" when they cover it. Ask students to use this information to predict the number of trapezoids and triangles to cover the shape and to share why they made their predictions. You also may want to ask them to tell how many blue rhombuses will cover the shape.

Although all students won't be able to figure out the number of triangles or rhombuses by thinking about the number of hexagons, keep encouraging such thinking. As discussed in the **Teacher Note,** Seeing Relationships Between Shapes (p. 27), students will use a variety of strategies in dealing with these problems. It may take a long time for some students to figure out, for example, that since 6 triangles make a hexagon, a shape made with 5 hexagons will need 30 triangles to cover it.

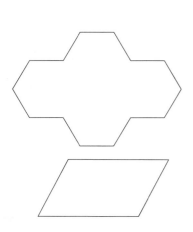

Shape E: Some students may initially think 5 hexagons will cover the shape, seeing 1 in the middle. See also comments for Shape D, above.

Shape F: After students explain that 6 blue rhombuses will cover the shape, ask them how many green triangles are needed. Ask if knowing the number of rhombuses helped them figure out the number of triangles.

After students have figured out how many triangles, ask how many trapezoids. This question attempts to get students to reverse the reasoning they have used in the past. Since it takes 3 triangles to make a trapezoid, and 12 triangles to cover the shape, they may see that it takes 4 trapezoids (in a sense, joining sets of 3 triangles to make trapezoids).

You may want to point out to students that Shape F is not a rhombus (since all sides are not the same length) and it is not a rectangle (since it doesn't have square corners). It is a four-sided shape called a parallelogram.

The student sheets for Shapes G and H ask students to make their own choices of blocks to use to cover their shapes. For each shape, start your discussion by asking:

Which pattern block is the easiest to predict for this shape? Why?

After students share their predictions and reasons, they check their predictions by covering the shape. Then they make predictions for other blocks. Compare students' written results with the predictions they have just made.

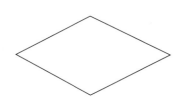

Shape G: As for Shape F, most students will make predictions for blue rhombuses and triangles. Ask students to discuss and check whether trapezoids also will cover the shape.

Ask students to describe Shape G. Some students may insist that it is a diamond (because of its orientation) and not a rhombus (even though the blue rhombus pattern block is similar to it). Don't expect students to recognize and name all the shapes. It takes several years for students to develop an understanding of the attributes of different shapes and how they are named and classified.

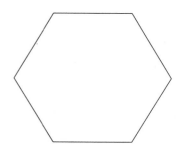

Shape H: Since the shape is a hexagon, many students will think it can be covered with hexagons. Some may remember that this was impossible. Ask students to discuss why.

Since this shape can't be covered with hexagons and is the most open, it is the hardest to make an initial prediction for. When students have found that 8 trapezoids will cover it, ask them to predict the number of triangles, and perhaps the number of blue rhombuses.

You might want to compare the results for Shape H and Shape E. How can students explain that they are different shapes yet use the same number of trapezoids, triangles, and rhombuses?

Sessions 4 and 5 Follow-Up

Composing New Shapes with 2 Triangles   Students cut the squares on Student Sheet 10, Squares, into two triangles. They use each pair of triangles to make a new shape. The pairs of triangles must touch along at least part of one side, but they cannot overlap. Students trace at least four of the shapes they make on Student Sheet 11, Composing New Shapes with 2 Triangles. Remind students to save their triangles; they will use these shapes again for another homework assignment.

🏠 **Homework**

Gather students around the largest computer display you have to introduce the program. If your display is small, you may want to introduce the software to smaller groups of students over several days. Tell them that they will build pictures and designs with *Shapes* just like they do with pattern blocks. Demonstrate the following on the computer:

- How to open *Shapes* by double-clicking on the icon.

- How to open the Shapes Pictures activity by clicking on it once.

- How to read the directions, then click on **[OK]** (or press **<return>**).

- How to get several shapes from the *Shapes* window.

- How to slide shapes by dragging them (point out that they snap into position when their sides are close).

- How to turn shapes with the two Turn tools.

- How to use the Erase One tool to erase one shape at a time and the Erase All button to erase all the shapes.

Note: For complete instructions, see the *Shapes* Teacher Tutorial (p. 129).

Working in pairs, students use Free Explore in the *Shapes* software to make their own pictures or designs.

Many students who are using computers and the *Shapes* software for the first time will need assistance. Most of the time, their questions will require short answers or demonstrations. (See the Teacher Tutorial on p. 129.) You do not have to be the only source of help for these students. For example, a student may not be aware of how to use the mouse. Often students who are more familiar with computers can assist those who need help.

As you observe students using the Shapes Pictures activity, ask them to describe what they are doing. Talk to the students about the way they are moving the blocks (sliding the blocks, and using the Turn and/or Flip tools). This will help them become more aware of these geometric motions. Just as important, it will help them become familiar with seeing shapes in different orientations and realizing that changing the orientation does not affect the shape's name (class) or attributes.

If students indicate the need for them, introduce these tools:

- The Flip tools flip, or reflect, shapes over a vertical or horizontal line. If you click on the shape with the first flip tool, the shape flips over a vertical line through the center of the shape.

- The Duplicate tool makes copies of shapes.

- The Arrow tool selects shapes. This is useful if students wish to apply a tool or command such as **Duplicate** or **Bring to Front** to several shapes at the same time.

- The Magnification tools allow you to make shapes bigger or smaller. *Shapes that are different sizes will not snap to each other.*

- The Glue tool glues several shapes together into a "group," a new composite shape that can be slid, turned, and flipped as a unit.

- The Hammer breaks apart a glued group with one click. If you hammer a shape (for example, blue rhombus) that is not part of a group, it will break a single shape into smaller shapes (for example, 2 green triangles).

If you wish to discuss students' work later, have them save their pictures on disk.

Managing the Computer Activities

The grade 2 *Investigations* curriculum uses two software programs developed especially for the curriculum. *Shapes* is introduced in *Mathematical Thinking at Grade 2* and used in *Shapes, Halves, and Symmetry.* *Geo-Logo* is introduced in *How Long? How Far?* Although the software is included in only these units, we recommend that students use the programs throughout the year. As students use the activities again and again, they develop skills and insights into important mathematical ideas.

How you use the computer activities in your classroom will depend on the number of computers you have available. Although we have included Free Explore in Choice Time activities, your computer setup may not be realistic for student use of computers during math class. If you have a computer lab available once a week or if you have only one or two computers in your classroom, you may want to schedule student use of computers throughout the day.

Regardless of the number of computers you can use, let students work in pairs on the computer. Working in pairs not only maximizes computer resources but also encourages students to consult, monitor, and teach one another. Generally, more than two students at one computer is difficult to manage; in most such cases, one or several students will end up having limited experience with the machine and the activity. But if you have an odd number of students, you can form one threesome.

Computer Lab. If you have a computer laboratory that has one computer for each pair of students, let all the students do the computer activities at the same time. During Choice Time, students will be able to work on the other choices. Plan to have students use the computer lab for one or two periods a week.

Three to Six Computers. The curriculum is written for this case and in many ways, it is the simplest to coordinate. If you have several computers in your classroom, you can use computer activities as a

Choice Time activity. You might introduce the computer and software to the whole class, using a large-screen monitor or projection device, or to small groups gathered around a machine. Then pairs of students can cycle through the computers, just as they cycle through other choices. Each pair should spend 15 to 20 minutes at the computer in one session. It is important that every student get a chance to use the computers, so you may have to allow students to use the computers at other times of the day. Monitor computer use carefully to ensure access for all students.

One or Two Computers. If you have only one or two computers in your classroom, students will definitely need to use the computers throughout the school day so that every pair of students has sufficient opportunity to do the computer activities.

Students who are not too familiar with computers may need help. Encourage students to work together and experiment and to see if they can figure out what they need to do and then to share what they've discovered with other students and with you. It is not unusual for students to discover things about the software that the teacher doesn't know.

Saving Student Work Students can save their work on the computer in two ways: on the computer's internal drive or on a disk. Instructions for saving work are on p. 157 and p. 165 of the *Shapes* Teacher Tutorial.

Building Buildings

Materials

- Computers with *Shapes* installed (optional)
- Student Sheets 3–6 (1 shape per pair)
- Pattern blocks (1 tub per 6–8 students)
- Interlocking cubes (60 per pair in bags or tubs)
- Student Sheet 12 (1 per student)
- Overhead projector (optional)
- Transparent pattern blocks (optional)
- Transparency of Student Sheet 6 (optional)
- Transparency of Student Sheet 12 (optional)
- Choice Time materials from previous sessions
- Student Sheet 13 (1 per student, homework)

What Happens

Students are introduced to two additional Choice Time activities. In The Last Block Game, students take turns covering shapes with pattern blocks. In Build a Building, they use interlocking cubes to build and count the number of "rooms" in multistory buildings. Students also continue work on other Choice Time activities. At the end of Session 8, they look for numerical patterns in their buildings. Their work focuses on:

- using logical reasoning and spatial relationships
- using multiples of a number to describe the structure of rectangular prisms

Start-Up

Quick Images   Flash images of arrangements of dots on the overhead. Ask students to draw the dots on blank paper after seeing the image the first time for 3 seconds. When students have finished drawing, show the image again for 3 seconds and tell them they can redraw the arrangement if they wish. Then display the image and ask: **Is your drawing the same as this picture of dots? If not, how is it different?**

Today's Number   Sometime during the school day, students brainstorm ways to express Today's Number. Suggest that students use three addends in each number sentence they write. Add a card to the class counting strip and fill in the next number on the blank 200 chart.

Activity

The Last Block Game

Two new activities are added to the list of Choice Time activities. The Last Block Game uses Student Sheets 3–6, which were also used for Predict and Cover. This game can also be played on the computer using any of the outlines in the Solve Puzzles activity of the *Shapes* software.

Teach the game by having two students play it as others watch. You may want to use Shape H for this introduction. If you have a set of transparent pattern blocks, students can play the game on the overhead using the blocks and a transparency of Student Sheet 6. If not, the pair can play the game with the class gathered around. Or show the game on the computer.

In The Last Block Game, students take turns placing a pattern block in a shape. The winner is the player who places the last block. As you teach the game, make sure that students understand where they can place a block. A block cannot be placed just anywhere inside the shape. Blocks must be placed so that when the game ends, the shape is completely covered with blocks—with no gaps or holes.

The Last Block Game is played with any of the shapes on Student Sheets 3–6 or on the computer. Use these pattern blocks to play: hexagons, trapezoids, blue rhombuses, and triangles.

The first player chooses a pattern block and places it in a corner of the shape. One side of the block must align with a side of the shape. (The piece doesn't have to fill a corner, but its edge must extend to the corner.)

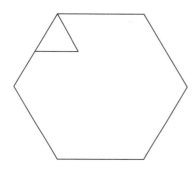

The second player chooses a pattern block and places it so it has a side in common with the first block. Players continue to take turns, each time placing a block next to at least one block that has already been placed. Once a block has been placed, it cannot be moved. The player who places the last pattern block in the shape wins.

In another version, the winner is the player who does *not* place the last block in the shape. As the available space within the shape grows smaller, players must think about the relationships among the pattern blocks to avoid placing the last block.

Activity

Cube Buildings

To introduce this activity, distribute interlocking cubes to students.

In this activity, you will use interlocking cubes to build buildings. Pretend that each cube is a room in a building. You can make a building that is one story (or floor), two stories, or more than two stories tall. There is one rule: Each story must have the same number of rooms and fit exactly over the one below it. Start out using the cubes to make a building that has three rooms and one story. When you have finished, place it in front of you.

Compare students' buildings. Most students will have made the first building below; a few may have made the second.

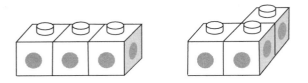

Now make your building taller—two, three, or even four stories. Remember, every story must have the same number of rooms and fit exactly over the one below it. How many rooms are in your building?

Students share their buildings and the number of rooms in them. Ask if students can predict the number of rooms that would be in their building if it were ten stories high.

Distribute Student Sheet 12, Build a Building, to each student. Explain that the rectangles lettered A, B, C, D, and E show the outline of the bottom story of five different buildings. Students use interlocking cubes to build each building. At the bottom of the sheet, they record the number of rooms in each building, and whether the building is one story, two stories, and so on. Point out that the last column asks for the number of rooms in ten-story buildings. (You may want to demonstrate how to record on the student sheet using the overhead projector.) When students are finished, they can make a building of their own.

Note: Student Sheet 12 is designed for use with ¾" interlocking cubes. If you have 2-cm cubes, you will need to redraw the outlines (or adjust your copier) for this student sheet.

Students can keep the student sheet in their math folders until they are ready to use it.

Activity

Choice Time

Add The Last Block Game and Build a Building to the list of Choice Time activities. Cross out Predict and Cover since students have completed it. If necessary, refer to Build the Geoblock (p. 19), and Solve Puzzles (p. 32) for materials and setup information and suggestions on what to watch for as you observe students working.

> 1. Predict and Cover
>
> 2. Build the Geoblock
>
> 3. Solve Puzzles (computer)
>
> 4. The Last Block Game
>
> 5. Build a Building

Choice 4: The Last Block Game

Materials: A shape from Student Sheets 3–6, Predict and Cover, and pattern blocks; or the Solve Puzzle activity on the *Shapes* software

This game can be played on or off the computer. Students take turns placing a pattern block in an outline. The person who places the last block wins.

Choice 5: Build a Building

Materials: Interlocking cubes; Student Sheet 12, Build a Building

Students use interlocking cubes to build different buildings. They record the number of rooms in their buildings on the student sheet.

Observing the Students

Use the following notes and questions as a guide for observing students.

The Last Block Game

Students have the opportunity to use strategy, although many students enjoy playing the game without much strategy. If students use some of the less-open shapes, such as the first four shapes on the student sheets, there is more possibility that students will see playing options. However, students may not find these games as interesting as using some of the other shapes. Notice when students see and take advantage of strategic moves.

- Do students start planning a strategic move several moves before a last block is placed or immediately before?
- Do students consider what could happen in subsequent turns if they place a particular block in a particular location now?

Build a Building

How do students build the buildings? See the **Teacher Note,** Students' Thinking About Build a Building (p. 43), for examples of different students' thinking.

Class Discussion: Build a Building

Allow about 20 minutes at the end of Session 8 for a class discussion. Students should have Student Sheet 12 in their folders or face down. Distribute interlocking cubes and have students build a building that has five rooms on each floor and is five stories tall. (If you don't have enough cubes, students can work in pairs.) Have students share how many cubes are in the building and how they figured it out.

Next, ask students to take out Student Sheet 12, Build a Building, and share their findings for Building A. As they state their results, record the numbers, on the board: 5, 10, 15, 20, 25, . . . 50.

Look at these numbers. What do you notice?

Students may point out that they are counting by 5's or skip counting. Ask them if they can explain why they got the 5's when they reported on the number of rooms in the building.

Ask students to share results for Buildings B and C. Why do the numbers come out the same? Ask a student to build a three-story building for Building B and another student to build a three-story building for Building C and compare the two buildings.

Sessions 6, 7, and 8 Follow-Up

🏠 Homework

Composing New Shapes with 3 or 4 Triangles   Students use 3 or 4 of the triangles they cut out for the previous homework to make new shapes. They trace at least 3 new shapes on Student Sheet 13, Composing New Shapes with 3 or 4 Triangles.

🔷 Extension

Guess My Building   When most students have completed Build a Building and are ready for a challenge, introduce the game Guess My Building. Two pairs of students play together. Each pair uses cubes to build a building. A building may have from 1 to 5 stories, with 1 to 10 rooms in each story. Partners hide their building so it is not visible to the other pair. The first pair describes their building in one of two ways.

1. Our building has ____ rooms on each floor. It is ____ stories high. How many rooms does our building have?

2. Our building is ____ stories high. It has ____ rooms altogether. How many rooms are there on each floor?

The second pair builds the first pair's building and answers their question. Pairs then reverse roles. The second pair describes their building, and the first pair builds it and answers their question.

Students' Thinking About Build a Building

Students may use a wide variety of strategies in solving the Build-a-Building problems. Many students may solve these problems by detecting numerical patterns. That is, they see that the answers are successive numbers in a skip-counting sequence. They often don't examine the cube buildings very carefully, relying solely on these number patterns.

Some students may simply add the number of cubes in each successive story. For example, to find the number of rooms in the fourth story of Building A (see Student Sheet 12), they might simply add 5 to 15 because they have already determined that there are 15 rooms in 3 stories.

Other students, however, may count *all* the rooms whenever a new story is added. Their counting may be organized by stories, columns, or neither. For example, we have seen students count the rooms in Building C in the following ways:

Building C:

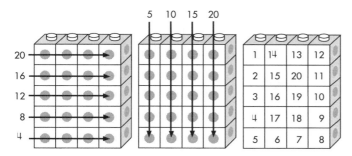

Many students have difficulty visualizing the organization of cubes within the buildings. This often makes their enumeration of cubes erratic. For example, Ayaz correctly counted the rooms in the first story of Building B, but had difficulties with the second story. For the two-story building, he counted the tops of the cubes in the second story but counted the sides of the cubes in the first story.

Building B:

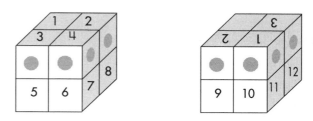

You will see a variety of enumeration strategies used by students—counting by 1's, skip counting, using addition. The strategy used by a student often depends on how well he or she knows particular skip-counting sequences.

Finally, you will also see a number of student difficulties with counting. One of the most common is attempting to skip count but losing track of the number of times one has counted. For example, after having found the number of rooms in a 5-story building with 3 rooms per story, Temara was trying to predict the number of cubes in a 10-story building with 3 rooms per story.

Temara counted on from 15 by groups of 3, tapping the top 3 cubes in the fifth story successively: 16, 17, 18; 19, 20, 21; 22, 23, 24; 25, pause. Temara stopped, saying that she forgot how many stories she had counted. She then repeated this procedure, this time holding up one finger every time she counted 3 cubes. She stopped this counting when she had extended 5 fingers, giving a correct answer of 30 rooms.

What Is a Rectangle?

What Happens

Session 1: Investigating Quadrilaterals   Students play Guess My Shape Rule and sort the Shape Cards by the number of sides. They discuss whether all the three-sided shapes are triangles. Students sort the shapes with four sides in different ways. They write an answer to the question: What is a rectangle?

Session 2: Which Rectangle Is Biggest?   Students cut out and arrange rectangles in order from biggest to smallest. They discuss how they determined the order and whether the rectangles can be ordered a different way. Students cover the rectangles with tiles to determine which takes the most squares to cover.

Session 3: Building Rectangles   Students use color tiles to make rectangles and describe and draw what they've made. They investigate the number of different rectangles that can be made from a given number of tiles. They continue each of these activities during Choice Time.

Sessions 4 and 5: Describing Rectangles Students draw rectangles seen as Quick Images and share how they saw them in their minds. They are introduced to Growing Rectangles, a computer activity in which they build rectangles by gluing and duplicating rows or columns. As part of their continuing Choice Time, students build rectangles described in riddles and write their own rectangle riddles.

Session 6: Picturing Rectangles   Students build rectangles to match each other's descriptions, then discuss what makes a good description of a rectangle. As an assessment, students draw a picture of a rectangle they are shown and write a description of it.

Mathematical Emphasis

- Identifying triangles and rectangles based on the number of sides, the number of corners, and the number of square corners
- Visualizing, constructing, and drawing rectangular arrays
- Using numbers to compare rectangular arrays

What to Plan Ahead of Time

Materials

- Prepared set of Shape Cards from Investigation 1 (Session 1)
- Yarn or string (Session 1)
- Index cards: 2 (Session 1)
- Overhead projector (Sessions 1, 3–6, optional)
- Scissors (Session 2)
- Color tiles: 1 tub per 6–8 students (Sessions 2–6)
- Plain paper: 1–2 sheets per student (Sessions 2–3, 6)
- Chart paper (Session 3, optional)
- Construction paper: 2–3 sheets per pair (Session 3)
- Tape and paste or glue sticks (Session 3)
- Computers: Macintosh II or above, with 4 MB of internal memory (RAM) and Apple System Software 7.0 or later: 1 for every 4–6 students (Sessions 4–5, optional)
- Apple Macintosh disk, *Shapes* (Sessions 4–5, optional)
- Projection device or large-screen monitor on one computer for whole-class viewing (Sessions 4–5, optional)

Other Preparation

- Duplicate student sheets and teaching resources, located at the end of this unit, in the following quantities. If you have Student Activity Booklets, copy only the transparencies marked with an asterisk.

For Session 1

Student Sheet 14, Looking for Quadrilaterals (4-Sided Figures) (p. 183): 1 per student (homework)

For Session 2

Which Is Biggest? (p. 190): 1 of each sheet per pair of students

Student Sheet 15, Which One Has the Most? (p. 184): 1 per student

For Session 3

Student Sheet 16, Inch Graph Paper (p. 185): about 5 per student (class), plus 1 per student (homework)

Student Sheet 17, Only One Rectangle (p. 186): 1 per student (homework)

For Sessions 4–5

Quick Images: Rectangular Arrays* (p. 203): 1 transparency if you have not already done so. Cut it apart into individual images.

Student Sheet 18, Growing Rectangles (p. 187): 1 per student if you are using the computer activities

Student Sheets 19–20, Rectangle Riddles (pp. 188–189): 1 of each sheet per student

Student Sheet 16, Inch Graph Paper (p. 185): several per student (homework)

- If you are using computers with this unit, try the Create a Tiling activity on the *Shapes* software before demonstrating it for students. See the Teacher Tutorial (p. 129) for additional information on the software. The **Teacher Note**, Introducing the *Shapes* Software (p. 36), provides suggestions on introducing the software to your class. (Sessions 4–5)

Investigating Quadrilaterals

Materials

- Prepared sets of Shape Cards (from Investigation 1)
- Yarn or string
- Index cards (2)
- Student Sheet 14 (1 per student, homework)

What Happens

Students play Guess My Shape Rule and sort the Shape Cards by the number of sides. They discuss whether all the three-sided shapes are triangles. Students sort the shapes with four sides in different ways. They write an answer to the question: What is a rectangle? Their work focuses on:

- sorting shapes according to the number of sides
- sorting quadrilaterals in different ways
- identifying rectangles

Start-Up

Quick Images Using the 10 Frames images, flash a 10 Frames number, such as 5, on the overhead. Have students share how many dots there are and how they know. Flash several other 10 Frames images. After each one, students share how they found the number of dots. Encourage comments, such as, "I knew it was 9 because there was only 1 dot missing" or "It's 6 because the top row and 1 more were filled." For complete details on this routine, see p. 125.

Today's Number Sometime during the school day, students brainstorm ways to express Today's Number. Suggest that students use subtraction in each number sentence they write if they haven't already done so. Add a card to the class counting strip and fill in the next number on the blank 200 chart. For complete details on this routine, see p. 116.

Activity

Guess My Shape Rule

Introduce Guess My Shape Rule. This game is similar to Guess My Rule in the unit *Mathematical Thinking at Grade 2,* in which the teacher thinks of different attributes to sort students in the class. In this game, one attribute that is common to some of the shape cards, such as four sides, is used to sort the Shape Cards into groups.

Gather students in a circle and arrange a set of Shape Cards face up.

We're going to use our Shape Cards again to play Guess My Shape Rule. I am thinking of a secret rule about the shape cards. Your job is to figure out what that rule is. Some shapes fit my rule, and some do not. If you think you know what my rule is, don't say it out loud.

Using the rule THREE SIDES, choose a few shapes that fit the rule and a few that do not. Place these shapes in the center of the circle. You can separate the two groups with yarn or string. Write labels for each group on index cards: FOLLOW MY RULE and DON'T FOLLOW MY RULE. Place the remaining Shape Cards in a pile and ask several students to choose a card.

Ebony and Tim, please place your Shape Cards on this side of the yarn. These two shapes follow my rule. Paul and Laura, please place your Shape Cards on the other side of the yarn. These shapes don't follow my rule.

If you think you have an idea about where your shape goes, raise your hand. Chen, do you think your shape follows my rule or doesn't follow my rule? Yes (or no), your shape follows (or doesn't follow) my rule. Please place your shape on the correct side of the yarn.

Emphasize the importance of all information—looking at shapes that do not fit the rule as well as those that do. Continue the clue gathering until many students have had a chance to place their shapes.

When most students seem to have a good idea of the rule, ask a volunteer to state the rule and give his or her reasons. Often, students have different ways of describing the same rule, such as triangles or three sides. Sometimes students will come up with a rule that fits the evidence but is not the rule you had in mind. If this happens, acknowledge the student's good thinking even though it did not lead to your secret rule.

❖ **Tip for the Linguistically Diverse Classroom** Ask students with limited English proficiency to point to the common attribute of each shape that they think is the basis of the rule. Identify the attribute orally. When English-proficient students are suggesting a rule, encourage them to also point to the attributes they are noting.

Discuss whether all the three-sided shapes are triangles. Students may disagree about this. For example, some may think that Shapes L and R are *not* triangles because they are long and skinny. Next, remove all triangles and set them aside.

Now we'll play Guess My Shape Rule without the triangles.

If students mention that all these shapes have four sides (except Shape I), tell them that four-sided shapes are called quadrilaterals. Don't insist that students learn the word, although some may enjoy adding it to their vocabulary.

Ask one or two students to pick a rule and whisper it in your ear. Play the game several times using their rules. Encourage discussion about whether shapes follow particular rules and whether the rules are clear enough so it can easily be determined whether shapes do or do not follow them.

If you haven't discussed which shapes are rectangles after several students have turns, state that you have another rule. This time place all the rectangles (Shapes A, F, G, M, and Q) by the label Follow My Rule.

Encourage students to discuss what's the same about all the shapes that follow your rule. Some students may call them rectangles; others may disagree, stating that some of them are not because they are squares or too long and skinny. Students may point out that they all have four straight lines, they are not slanty, or they all have four corners. You might have students look at the corners, asking whether the corners in the shapes that follow your rule are different from those that don't follow your rule.

After students have discussed the attributes of the shapes, tell them that mathematicians call all these shapes rectangles. Some students may disagree. See the **Teacher Note,** What's a Rectangle? (p. 50), for a description of second graders' perceptions of rectangles.

As students return to their seats, give everyone a sheet of paper.

Suppose you wanted to describe a rectangle to someone younger than you. Write what you would tell him or her. [*Write on the board, "What is a rectangle?"*]

This activity will provide you with information about students' current understanding of rectangles. Encourage them to write about what they think about rectangles. If students have difficulty getting started, ask them to tell you out loud and then to write what they said on paper. If some finish quickly with just a brief response, such as, "It's a shape," urge them to write more about what they know.

❖ **Tip for the Linguistically Diverse Classroom** To help students respond to the question, help them examine the attributes of a rectangle. For example, count the sides and corners and label them 1, 2, 3, 4. Discuss the length of the sides noting which are the same.

Students can save their writing in their math folders, or you may want to collect them and save them together so you can ask students to repeat the assignment later in the year and compare results.

Writing: What Is a Rectangle?

Session 1 Follow-Up

Looking for Quadrilaterals (4-Sided Figures) Students look for examples of rectangles and other quadrilaterals at home. They find things they can bring to school, such as used envelopes or postage stamps or cut pictures from newspapers, magazines, or flyers. They can draw things they can't bring in on Student Sheet 14, Looking for Quadrilaterals (4-Sided Figures).

🏠 **Homework**

Teacher Note ⟩ *What's a Rectangle?*

What's so hard about understanding what a rectangle is? The definition of a rectangle seems very clear to us as adults: It is a four-sided polygon with opposite sides parallel and four right angles. If we just show rectangles to students and describe what they are, won't they understand?

Research with students throughout the elementary grades makes it clear that understanding what a geometric object is and how it's classified does not come from learning a definition. Although students hear the definitions of rectangle and triangle all through the elementary grades, and can even repeat those definitions accurately, they still may not agree that an unusual rectangle (for example, a square) or an unusual triangle (for example, a scalene triangle) fits the definition. Definitions of everyday objects often depend on the perception of a typical representation.

For example, some fourth and fifth grade students do not believe that if a square is rotated 45°, it is still a square.

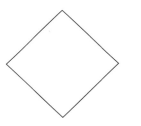

They have an implicit definition, based on seeing many squares presented, that a square must be "sitting" on a side in order to be a square. Similarly, students will argue that a square is not a rectangle—even though they know the definition of a rectangle—because their image of a rectangle is that one of its dimensions must be longer than the other.

In order for students to develop an understanding of a definition, they must examine many examples and counterexamples of the thing they are defining. They must describe, compare, and argue about what is similar and what is different. In order to develop definitions, students must develop the habit of paying careful attention to characteristics; careful observation and description are what we want students to begin to do in this unit.

Asking students to write about what they think a rectangle is provides you with a glimpse into their thinking. It is a record that you can save and perhaps compare with a similar writing assignment done later in the year.

Here are some examples that show the range of student ideas in a second grade classroom:

> **Naomi**
>
> A rectangl is a shape that has 4 sides. It is like a square but it is a little longer, it lookes like this peas of paper but a little smaler. A rectangl is long. Sometimes it is a neet shape to work with.

> **Tory**
>
> It looks like a square and it has four sids and four korners. A rectangel is vere long.

Continued on next page

Rosie

A rectangle has 4 sides. It look like a square but longer.

Chen

It does not have equal sides.
It is not a square.

A definition of a rectangle is offered by example in Investigation 2, based on the students' experience with sorting quadrilaterals. However, don't expect most students to finish this investigation with a complete understanding of what a rectangle is, or what is included in the definition and what isn't. Rather, as they are working with rectangular arrays to gain more experience with the characteristics of rectangles, expect them to continue considering new examples that arise: Could this be a rectangle? What are its characteristics? What would our definition have to say if this were included?

Which Rectangle Is Biggest?

Materials

- Color tiles (1 tub per 6–8 students)
- Scissors
- Which Is Biggest? sheets (1 of each per pair)
- Plain paper (1 sheet per student)
- Student Sheet 15 (1 per student)

What Happens

Students cut out and arrange rectangles in order from biggest to smallest. They discuss how they determined the order and whether the rectangles can be ordered a different way. Students cover the rectangles with tiles to determine which takes the most squares to cover. Their work focuses on:

- ordering rectangles
- defining "biggest" in different ways
- finding the number of tiles that cover rectangles

Start-Up

Today's Number   Sometime during the school day, students brainstorm ways to express Today's Number. Suggest that students use both addition and subtraction in each number sentence they write. Add a card to the class counting strip and fill in the next number on the 200 chart.

Homework   Students share examples of rectangles or quadrilaterals they found at home.

Activity

Ordering Rectangles

Distribute scissors and the Which Is Biggest? sheets to each pair of students and have them cut the rectangles apart. Provide each student with paper.

Which rectangle do you think is the biggest? Which is the smallest? Work with your partner to put them in order, starting with the rectangle you think is the biggest. Then on your paper, write the order you've used. When everyone has finished, we'll discuss what you have done.

Students may use any materials they want to determine the order of the rectangles. Observe what students do and if there is any discussion or disagreement among students about how to determine the size of the rectangles. (If some students are extremely slow cutting out the rectangles, you may want to ask others to help them so that all have enough time to complete the activity.)

When all students have ordered the rectangles, call the class together. You may want to seat students in a circle, so all can see a set of rectangles. Students should bring their rectangles and paper with them.

Ask one pair to share their order, putting the rectangles in the center for all to see. Most second graders will have arranged them beginning with the tallest rectangle, and when two rectangles are the same height, they put the wider one first. A common order is: C, F, B, A, D, G, E, with each rectangle "standing up." When one student has shared an order, ask if someone has a different order.

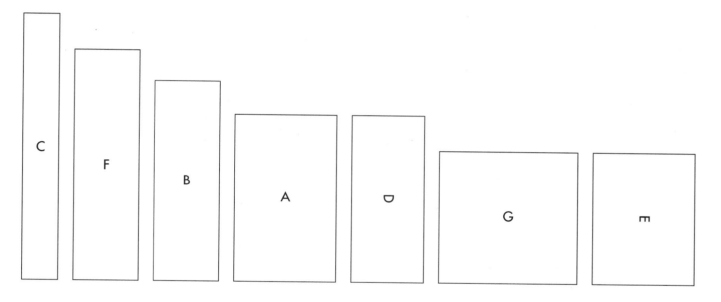

A few students may have arranged them differently, for example, "lying down" or by area. Discuss with students how to define "biggest." Encourage them to share their thinking. Students should explain their arrangements, from biggest to smallest, until all orders have been shared.

If no one has a different order, take the longest rectangle (C), and turn it the other direction.

You have said that rectangle C is the biggest because it's the tallest. If I turn it this way [*so the long direction is horizontal*], **is it still the biggest rectangle? Why or why not?**

Students discuss their opinions. Some students may state, and not see any contradiction, that rectangle C is the biggest when it is standing up and the smallest when it is lying down. If students think that the order changes when rectangles are turned a different direction, ask them if all were turned, which would be the biggest and what would be the order. You also might ask students whether two different rectangles can both be "the biggest."

Covering Rectangles

With rectangles still visible from the previous activity, pose the following question.

Suppose these rectangles are chocolate bars. Which one would have the most chocolate? Which would have the least? How could you find out?

❖ **Tip for the Linguistically Diverse Classroom**   To help students with limited English proficiency imagine the chocolate bars, show them a candy bar or candy bar wrapper and discuss the shape.

Students may offer suggestions, such as using a ruler. If no one mentions it, suggest covering the rectangles with color tiles. Each tile can represent a square of chocolate. To demonstrate, have a student cover one rectangle, then ask students to find how many tiles would cover each of the other rectangles.

Working individually or in pairs, students order the rectangles by covering them with tiles. They record the number of tiles to cover each of the seven rectangles on a separate piece of paper. When they have finished, ask each student to complete Student Sheet 15, Which One Has the Most?

Call students together to share their results. Do they all agree? If not, cover the rectangles with tiles. Discuss those that take the same number of tiles to cover.

Look at rectangles B and E. Which has more chocolate? Is one bigger than the other?

Encourage students to discuss whether they think these two shapes are the same size or different sizes. Some students may be sure that the tallest rectangle is biggest. Others may change what they consider "biggest" depending on the attribute. This is fine, since the meaning of "biggest" depends on the attributes being compared and the context of the situation. For example, when comparing pencils or fishing poles, we focus on the attribute of length, so in this context "biggest" means longest. To find which of several rooms will use the most carpeting, we are interested in which floor has the largest area.

Building Rectangles

What Happens

Students use color tiles to make rectangles and describe and draw what they've made. They investigate the number of different rectangles that can be made from a given number of tiles. They continue each of these activities during Choice Time. Their work focuses on:

- ordering rectangles
- finding different rectangular arrays using the same number of tiles
- describing ways of arranging square tiles in rectangular arrays
- representing arrays by drawing on squared paper

Start-Up

Today's Number   Sometime during the school day, students brainstorm ways to express Today's Number. Suggest that students use both addition and subtraction in each number sentence they write. Add a card to the class counting strip and fill in the next number on the blank 200 chart. For complete details on this routine, see p. 116.

Materials

- Color tiles (12 per student)
- Overhead projector (optional)
- Student Sheet 16 (4–5 per student, class; 1 per student, homework)
- Construction paper (2–3 sheets per pair)
- Tape
- Paste or glue sticks
- Plain paper (2 sheets per student)
- Chart paper (optional)
- Student Sheet 17 (1 per student, homework)

Activity

Building Tile Rectangles

Distribute color tiles and two sheets of plain paper to each student.

We are going to put color tiles together to make our own rectangles. Take six tiles and make a rectangle. Then cover it with a sheet of paper.

When everyone has made a rectangle, give the next direction.

Take a quick peek at your rectangle and cover it up again. Then draw a picture of it on the other sheet of paper. Show all six tiles in your drawing.

Now take the cover off your tiles. Compare your drawing to the tiles. If you need to make changes, draw a new picture beside your first one. This time you don't need to cover your tiles.

Who can describe the rectangle you've made so others can tell if they've made the same one?

As a volunteer describes a rectangle, build it on the overhead projector.

Do you all think this is the rectangle Tim made? Why or why not?

If the description is not clear, encourage the student to tell you more about the rectangle. When the rectangle is built on the overhead, ask the class if anyone has made the same rectangle.

Can anyone describe the rectangle in a different way?

As students describe the same rectangle in a variety of ways, record their descriptions on the chalkboard.

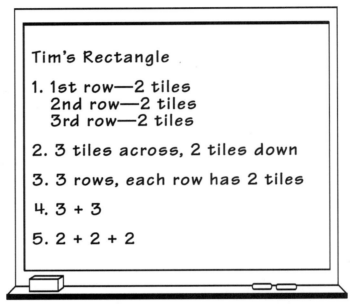

Tim's Rectangle

1. 1st row—2 tiles
   2nd row—2 tiles
   3rd row—2 tiles

2. 3 tiles across, 2 tiles down

3. 3 rows, each row has 2 tiles

4. 3 + 3

5. 2 + 2 + 2

Many of you made the same rectangle Tim made. Did anyone make a different rectangle? Can someone describe the rectangle he or she made?

Again, construct the rectangle on the overhead from the student's description. Then discuss whether the rectangle is different from the first. Encourage others who have made a rectangle different from these to describe their work.

There are two rectangles students can make using six tiles, a 1-by-6 rectangle and a 2-by-3 rectangle. It is common for students to see two rectangles with the same dimensions as different, if one is rotated 90°.

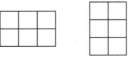

If students do not recognize these as the same rectangle, but do say they are turned in different directions, ask if the rectangles would be the same if one was turned around. Introduce the term *congruent*, but you should not expect students to memorize the word and its meaning.

When two shapes are the same size and shape, we say they are congruent. It doesn't matter if they are turned in different directions. They are congruent if one shape would fit exactly on top of the other one.

Some students may make other shapes. If they do, ask them to point out the rectangles in the shapes they make.

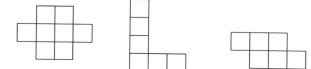

Yes, I can see that rectangle in your shape. But for this activity the whole shape made with the tiles must be a rectangle.

Some students may say that a rectangle one tile wide is not a rectangle, "It's a line." Ask, **What shape would it be if we drew around the outside edges and then took away the tiles?**

How Many Rectangles?

Each student takes 12 tiles. (If you think that students will focus on color arrangements of tiles, ask them to take tiles of the same color.)

How many different rectangles do you think can be made from 12 tiles? Make as many rectangles as you can. Then compare your work with a partner to see if you have each made any different ones.

As a class, discuss the rectangles students made. As students describe a rectangle, draw it on the overhead or on chart paper. Next to each one, record the ways students described their rectangles.

You have made one rectangle that has 2 tiles in each of 6 rows and another that is 3 rows with 4 tiles in each row. Do you think you've made all the possible rectangles using 12 tiles? See if you can make a different rectangle.

Students share findings. If students don't suggest a 1-by-12 rectangle, ask them if they can make a rectangle that uses 12 tiles and is only 1 tile wide.

You have just made two different rectangles using 6 tiles and three different rectangles using 12 tiles. There are some numbers that will make only one rectangle. Some numbers, like 12, will make three or more rectangles.

Tell students that investigating rectangles for some of these numbers will be one of their Choice Time activities.

Choice Time

Introduce and post a list of the first two Choice Time activities.

> 1. Describing Rectangles
>
> 2. How Many Rectangles?

Choice 1: Describing Rectangles

Materials: Color tiles; Student Sheet 16, Inch Graph Paper

Students use color tiles and make several rectangles. They choose one rectangle and draw it on the bottom half of Student Sheet 16, Inch Graph Paper, staying on the dotted lines. They fold the top half of the student sheet over their drawing and write a description (and/or draw a picture) of their rectangle on the top flap so that someone else can reproduce it without looking. Students will use their completed descriptions in Session 6.

Note: Leave the number of tiles in each rectangle up to students. If they build many large ones, tell them that the one they choose to draw must fit on the bottom of the student sheet.

Choice 2: How Many Rectangles?

Materials: Color tiles; Student Sheet 16, Inch Graph Paper; construction paper; tape; glue or paste

Students work with partners and use color tiles to investigate the number of rectangles they can make for one or two of the following numbers: 16, 18, 24, 36. When they think they have found all the rectangles for one of the numbers, students trace around them on the graph paper, cut them out, and paste them on a sheet of construction paper. Students may need to tape two sheets of graph paper together to record some of their rectangles.

When students have made rectangles for one or two of the numbers, they can choose a number of their own from 8 to 36 and see how many different rectangles they can make.

Observing the Students

Observe students as they work. Use the following suggestions of what to watch for.

Describing Rectangles

■ How do students build rectangles? Do they place one tile at a time until they have formed a rectangle, do they grab a handful of tiles and try to arrange them into a rectangle, or do they start with a specific number of tiles? Do they frame the outside first and fill in, or do they go by rows or columns? Do students choose small or large rectangles to build?

■ Are students able to copy their rectangles onto the graph paper? If they have difficulty copying their rectangle, suggest they place the tiles on the graph paper and trace around it.

■ How do students describe their rectangles in writing? If students have trouble getting started, suggest they first tell you how to build their rectangles. When students have completed a description, try building the rectangle without looking at the drawing to see if the description is clear.

My rectigl has two rows. It has four tiles in the first row and four in the secdend.

Inch Graph Paper
Student Sheet 16

How Many Rectangles?

■ How do students try to construct rectangles from a given number of tiles? Do they randomly try to arrange them, or do they structure them as an array in rows and columns? Do students see pairs of rectangles (for example, 6-by-3 and 3-by-6) as the same rectangle or different? See the **Dialogue Box**, Is It the Same Rectangle? (p. 60), for some typical second graders' thoughts about rectangle pairs.

■ Do students have a strategy for finding different rectangles for the same number? Or do they randomly try to form rectangles? Students will vary tremendously in their approaches to this task. Some students, as in the **Dialogue Box**, Strategies for Building Rectangles (p. 61), become quite sophisticated in figuring out how to make different rectangles with the same number of tiles.

Session 3 Follow-Up

Only One Rectangle Students cut out the 12 1" squares on Student Sheet 17, Only One Rectangle. Students use the squares (or Student Sheet 16, Inch Graph Paper, if they like) to figure out which numbers between 1 and 12 make only one rectangle. They record the numbers they find on Student Sheet 17 and write about what they noticed.

⌂ **Homework**

◼D◼I◼A◼L◼O◼G◼U◼E◼ ◼B◼O◼X◼

Is It the Same Rectangle?

During the Choice Time activity How Many Rectangles? (p. 58), Franco and Tim are building rectangles with 18 tiles, trying to find all possibilities. As they build, the issue of congruence comes up: Is a rectangle built "the 9 way" (9 across by 2 down) the same as a rectangle built "the 2 way" (2 across by 9 down)? As the dialogue opens, Tim is describing the rectangles to his teacher, who is circulating among the students, when Franco announces he's found another way to make a rectangle with 18 tiles.

Tim: This one's long; it's from 1's. This one's 2's. It's 9 long. And we made one from threes and it's 6 long. And we're going to make one more—

Franco: I found another one! Do it the 9 way!

Tim: We already did it the 2 way.

Franco: But if you do it the 9 way, it will make another rectangle. [*He demonstrates with a 3-by-6 rectangle, displayed so that it is 3 across.*]

Franco: So now we change this one this way: 6 + 6 + 6.

Tim: We already did that, though.

Franco: But this is the 6 way. [*He colors the 6-by-3 rectangle as he talks.*]

Tim: But it's 6 on *this* side.

Franco: But that's the 3 way.

Franco brings his new 6-by-3 rectangle to the display.

What will happen if you put it on top of this rectangle [*the 3-by-6*]? Why don't you try it?

Franco: It's the same. But if I turn it . . .

Does it make a difference if you turn it?

Franco: No, but that's the 3 way and this is the 6 way.

This discussion illustrates a common mathematical issue: When are mathematical objects mathematically equivalent and in what ways? For example, we agree that $3 \times 5 = 5 \times 3$. However, a game played with three teams of five people will not be the same as a game played with five teams of three people. A triangle and a rectangle are the same in that they are both closed straight-sided shapes—they are both polygons. However, they are quite distinct in other respects. What is the "same" depends on what attributes we attend to. Through discussions like these, students develop experience with making these kinds of distinctions. The 3-by-6 and 6-by-3 arrays are congruent; they are the same size and shape. But they can be different if we attend to position and orientation. Tim and Franco are thinking about this distinction.

Strategies for Building Rectangles

In this conversation during the Choice Time activity How Many Rectangles? (p. 58), students are sharing their strategies for finding all the rectangles for numbers they have chosen. Harris uses his knowledge of number to find rectangles for 30. This motivates some students to consider what happens when you split an array to make a new array. Bjorn uses a spatial model to do his splitting, while Lila uses a numerical one.

What strategies did you use to find all the rectangles for a number?

Harris: I started with one column of 30. Since 30 is an even number, you can split it. Since two of the same number gives you an even number.

Naomi: Since 30 is even, you can split it in two to get another way of making it.

What did you get when you split it?

Naomi: It was 15 down and 2 across.

Can you split that again?

Lila: Yes, but it's odd.

Naomi: Right, 15 is odd.

Bjorn: You can split that into 5's. You have two columns. Take 2 of the 5's off and then put them together to make 10 and you'll have 3 tens. [*He comes to board.*] I took these off to make their own row.

Bjorn took these top two groups of five off and moved them down next to the others to make another rectangle.

So this is 10 tiles high and how many across?

Bjorn: 3.

Can we break it down more?

Lila: You can make 15 by 7½'s.

Let's stick with whole numbers for now.

Paul: That's the limit.

Bjorn: Yes! Yes! Six 5's.

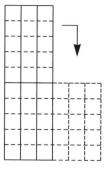

Bjorn took these top 3 groups of 5 and moved them down next to the bottom 3 groups of 5, going from a 3-by-10 to a 6-by-5.

Lila: A 6-by-5. 6 times 5.

How do you know?

Lila: Well, 5 + 5 = 10. 5 + 5 = 10. 10 + 10 = 20. And then the other 5 + 5 = 10 is 30.

Bjorn: I did it a different way from the way we did it before. We'd broken the 5's off, so we had 3's. I took the 5's off and put them down here.

Describing Rectangles

Materials

- Prepared set of Quick Images: Rectangular Arrays
- Color tiles (1 tub per group)
- Overhead projector (optional)
- Student Sheet 18 (1 per student if using computers)
- Computers with the *Shapes* software installed (optional)
- Student Sheets 19–20 (1 of each per student)
- Student Sheet 16 (several per student, homework)

What Happens

Students draw rectangles seen as Quick Images and share how they saw them in their minds. They are introduced to Growing Rectangles, a computer activity in which they build rectangles by gluing and duplicating rows or columns. As part of their continuing Choice Time, students build rectangles described in riddles and write their own rectangle riddles. Their work focuses on:

- visualizing and reproducing rectangles
- constructing rectangles by duplicating rows or columns
- using clues to construct rectangles

Start-Up

Today's Number   Sometime during the school day, students brainstorm ways to express Today's number using doubles in their number sentences. For example if the number they are working on is 92, possible combinations include: 46 + 46, 23 + 23 + 23 + 23, or 45 + 45 + 1 + 1. Add a card to the class counting strip and fill in the next number on the blank 200 chart. See p. 116 for complete details on this routine.

Activity

Quick Images: Rectangular Arrays

Give each group of students a bucket of color tiles. Ask students to empty the tiles on the table so they can be used quickly.

We are going to play Quick Images with rectangles. I will show you a rectangle on the overhead for 3 seconds. You will build the rectangle with color tiles. Then I will show you the rectangle again for another 3 seconds. This time you can check your rectangle and decide if you want to make any changes. When everyone is ready, I will turn on the overhead and you will compare your rectangle to the one on the overhead.

Place one of the prepared images, such as the 3-by-4 rectangle, on the overhead. Follow the Quick Images procedure. It's important to keep the picture up for as close to 3 seconds as possible. If you show the picture too long, students will build from the picture rather than their image of it; if you show it too briefly, they will not have time to form a mental image.

Then ask students to think about how to describe what they saw. They share these descriptions. Expect students to say things such as: "I counted 12 squares." "At first, I saw that there were 3 in a row, but on the second flash, I counted 4 in a row."

Continue the activity with one or two other rectangles. Open Session 5 using Quick Images: Rectangular Arrays.

Activity

On-Computer Activity: Introducing Growing Rectangles

If you are using computers in this unit, gather students around a large-screen display to introduce Create a Tiling activity in the *Shapes* software. You may want to distribute copies of Student Sheet 18, Growing Rectangles, so students can follow the directions while you demonstrate the activity.

We can use the computer to make rectangles in a way that we can't do with tiles—by gluing and duplicating, or copying, rows or columns. The rectangle we make on the computer will grow and grow.

Discuss the differences between rows and columns and have students decide whether to make a row or column for the demonstration. They also decide on the number of squares to use in the row or column.

Drag squares from the Shapes Window and place them together to make a row or column.

This is our first rectangle. It has only one row (column). How many squares in the rectangle?

Glue the squares together using the Glue tool. Then demonstrate how to use the Duplicate tool to make one copy of the row (column). Place the row (column) below (or beside) the first one.

This is our second rectangle. It has two rows (columns). How many squares are in the whole rectangle? How many squares will there be when we add a third row (column)?

After students predict, click once on the Pattern tool to make the third rectangle and check the number of squares. Predict for the fourth rectangle, then click again on the Pattern tool to add the row or column, and so on.

Students can record their predictions and results on Student Sheet 18 when they try this activity during Choice Time. Students can save Student Sheet 18 in their math folders until they are ready to use them.

Choice Time

Add the new Choice Time activities to the list. You may want to refer to the previous Choice Time (p. 58) for materials and setup information and suggestions on what to watch for as you observe students for Describing Rectangles and How Many Rectangles?

> 1. Describing Rectangles
>
> 2. How Many Rectangles?
>
> 3. Growing Rectangles (computer)
>
> 4. Rectangle Riddles

Choice 3: Growing Rectangles

Materials: Computers with *Shapes* installed; Student Sheet 18, Growing Rectangles

Students open the Create a Tiling activity on the computer and follow the directions on the student sheet to make a rectangle that "grows" by rows or columns. They record on the student sheet the number of squares in each rectangle.

Choice 4: Rectangle Riddles

Materials: Color tiles; Student Sheets 19 and 20, Rectangle Riddles

Students use color tiles and make a rectangle for each riddle. They draw the rectangle below the riddle. When they have finished, they write their own rectangle riddles.

Observing the Students

Growing Rectangles

Expect students to experiment with the activity before being ready to record a rectangle. Students find the Duplicate and Pattern tools quite exciting—they can make multiple copies with just a few clicks of the mouse. Once they have done some exploring, encourage them to make a growing rectangle.

- Are students able to follow the directions on the student sheet?
- Are students able to predict the number of squares in each of the rectangles before counting?

Rectangle Riddles

■ How do students use the clues in the riddles? Do they build first and then check the number of rows and columns? Do they use the clues independently or together? Do students realize that in riddle 3, just one piece of information is sufficient to build a rectangle? Do they have difficulty with other riddles, such as "This rectangle has 3 columns and 3 rows"?

■ How do students write their own riddles? Do they build a rectangle first? Do they provide sufficient information so someone else can build it?

Near the End of Each Session Inform students when 5 or 10 minutes remain in the session so they can finish and record their work. Have them clean up and put away materials, checking for stray tiles on the floor. Remind students to put their work in their folders and write in their Weekly Logs if they have not already done so.

Tell students that in the next session they will try to build each other's rectangles. You may want to collect students' written descriptions and drawings of rectangles from Describing Rectangles at the end of Session 5 so you can see how students are describing rectangles at this point, but don't make any corrections to what students have written.

Note: Save the descriptions students have written for Describing Rectangles for discussion during Session 6.

Sessions 4 and 5 Follow-Up

Making Rectangles Students take home several sheets of inch graph paper (Student Sheet 16) and try to draw rectangles for a number that hasn't been investigated by class members. You might want students to write the number they will investigate on one sheet of the graph paper before they go home.

 **Homework**

Fill the Rectangles Students use the *Shapes* software and open the Solve Puzzles activity. For puzzles 11–15, which are rectangles, they predict the number of squares to fill and try to figure out a quick and efficient way to fill them. They share their procedures for filling the rectangles.

Extension

Picturing Rectangles

Materials

- Color tiles
  (1 tub per group)
- Overhead projector
  (optional)
- Students' descriptions
  and drawings of rectangles (from Sessions 3–5)
- Plain paper (1 sheet per
  student)

What Happens

Students build rectangles to match each other's descriptions, then discuss what makes a good description of a rectangle. As an assessment, students draw a picture of a rectangle they are shown and write a description of it. Their work focuses on:

- following descriptions written by other students to build rectangles
- evaluating the effectiveness of written descriptions
- drawing a rectangle from memory and describing it in writing

Start-Up

Quick Images Repeat the activity Quick Images: Rectangular Arrays from Sessions 4 and 5. Show two or three images, different from those shown previously. See p. 125 for complete details on this routine.

Today's Number Sometime during the school day, students brainstorm ways to express Today's number. Add a card to the class counting strip and fill in the next number on the blank 200 chart. See p. 116 for complete details on this routine.

Activity

Describing Rectangles

Return students' written descriptions of rectangles, or have them take their descriptions out of their math folders. Distribute tubs of color tiles so that all students have easy access to them. Each student needs enough space to build the rectangles described. You may want students to work at their seats or on the floor.

Today we're going to think about what makes a good description of a rectangle. We want to know if the descriptions that were written are clear enough so that someone else can build the rectangle.

Some of you will read the description you wrote to the class without showing the picture. Then everyone will try to build your rectangle with color tiles. We'll compare the ones we've made with each other and with the one that the person who described the rectangle drew.

Choose several students to read their descriptions. After each description is read, the rest of the class uses color tiles to make the rectangle. Students should compare their rectangles with each other before the reader shows his or her drawing. If students have made different rectangles than the reader, encourage them to share why they think the rectangle they've made is the one that was described.

After several students have read their descriptions, discuss as a class what makes a description clear.

Who can share a description that they think was so clear that it was easy to build the rectangle?

Students might say things such as, "I knew how many tiles to use" or "It told me how many tiles to put in each row."

Assessment

Picturing a Rectangle

About 20 minutes before the end of Session 6, begin the assessment activity. Each student will need a sheet of paper and a pencil.

I'm going to make a rectangle in the center of the circle [*or on the overhead*]. Watch what I do. When I'm finished, I will cover the rectangle and you are to draw the rectangle I made.

Use color tiles to make a 3-by-7 rectangle. Build the rectangle, adding one tile at a time. When you have finished, tell students they have 3 seconds more to look at it. Then cover the rectangle, or turn the projector off.

Draw a picture of the rectangle. Then, below your drawing, write about it. To help you get started, imagine that you are talking to someone on the telephone, telling about your rectangle. What would you say so the person could draw the same rectangle?

Let students decide on their own how to draw their pictures in this activity, rather than showing them what to do. Students' drawings should represent their thinking about the organization of the square blocks within the rectangles. If you show them how to draw, their drawings may be rote and unconnected to their structuring of the arrays.

For samples and explanations of students' work on this task, see the **Teacher Note,** Assessment: Picturing Rectangles (p. 69).

Assessment: Picturing Rectangles

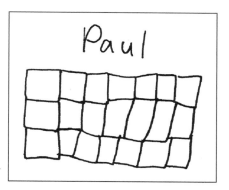

This investigation provides students with opportunities to investigate rectangular arrays. The array is a powerful mathematical model that can be used to think about addition, multiplication, fractions, and area. In this unit, students become familiar with the structure of an array, the way it is organized, and how to describe it both spatially and numerically.

You will find a wide variety of sophistication in students' structuring or organization of the rectangular arrays of squares. Some will not have the correct number of squares in their drawings. Some will have different numbers of squares in the rows (or columns). Some will have the number of squares in rows and columns correct but will not have the squares properly aligned.

Many students will draw square blocks one at a time. Others will draw the squares using full-length vertical and horizontal lines to make the rows and columns.

Looking at students' drawings and descriptions in this assessment will help you to understand how they are thinking about the structure of arrays and, in particular, whether they are seeing the array as a series of individual squares or whether they are beginning to organize these individual squares into larger chunks. You will be able to see some aspects of students' thinking from the finished products, but you will get even more information about their ideas as you watch them draw their arrays.

Here are samples of students' work from one second grade classroom. In this classroom, the teacher made the 3-by-7 rectangle on the floor with students gathered around. Some students drew it with 3 squares in a row; others with 3 squares in a column.

Paul constructed the array by using a series of single tiles. He first drew 3 squares in a column, then drew 6 more squares going across the top, 1 square at a time. He has a sense of the overall shape of a rectangles and tries to get his squares to fit into that shape, but he doesn't work with larger units—like a column of 3 or a row of 7.

Angel sees that there are 3 tiles going one way and 7 tiles going another way, but she has trouble coordinating these two. She builds a "frame" of 3 tiles going down and 7 tiles going across, then fills in that frame. She has a sense of the shape of a rectangle, knows that a rectangular array is made up of rows and columns, and draws her rows and columns so that they have the same number.

For some students, it is easier to describe an array than to draw it.

Continued on next page

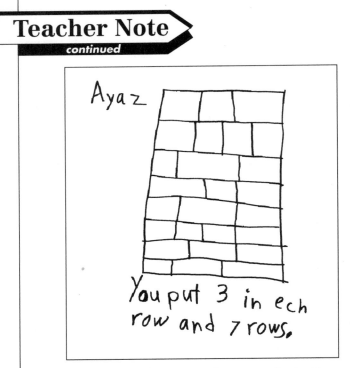

Ayaz describes the rectangle correctly, but he draws the rectangle a row at a time and doesn't pay attention to lining up the squares with those above. He also loses track of the number of rows and the number of squares in each row.

When Karina draws the rectangle, she draws each square separately, but she seems to have a complete picture in her head of how the squares are connected to make a rectangular array.

Several students seemed to view the array as a whole. They didn't need to draw individual squares but could divide up the rectangle using parallel lines.

Imani notices that there are columns but has a hard time coordinating the columns to make a rectangular shape.

Camilla started drawing the array with only two columns but realized she needed a third column and added it to her picture.

Continued on next page

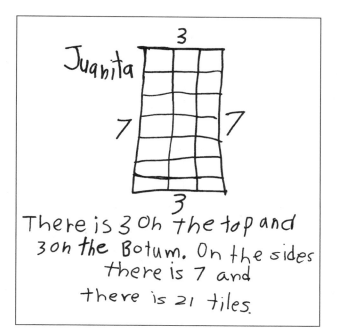

There is 3 0h the top and 3 0h the Botum. On the sides there is 7 and there is 21 tiles.

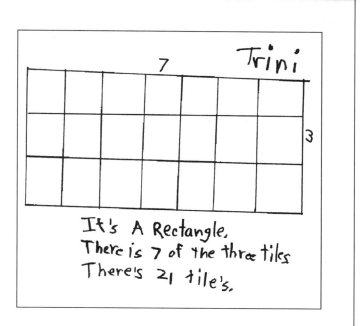

It's A Rectangle. There is 7 of the three tiles There's 21 tile's.

Juanita and Trini each drew a rectangle first, then divided it into rows and columns by using parallel lines. They both label their arrays in ways that show they see the coordination of the rows and columns to make a whole array.

Fractions of Geometric Shapes

What Happens

Sessions 1 and 2: Halves of Rectangles and Solids Students use two colors of tiles to make rectangular arrays that are half one color and half another. They use two colors to color the same rectangle in two ways—to show halves and not halves. They find Geoblocks that are one-half of another Geoblock. Choice Time is used as a Teacher Checkpoint.

Sessions 3, 4, and 5: Cutting Congruent Halves Students fold and cut different shapes into congruent halves, then make a shape of their own that can be cut and folded in half. Near the end of Session 5, they discuss which numbers will make half-and-half rectangles.

Session 6: Fraction Flags Students make rectangular flags that are divided into halves with each fractional part of the flag a different color. They decide on the size of their flag and how to divide it in half. Students use color tiles to design their flags, then copy the design by gluing construction paper squares on large paper.

Sessions 7 and 8 (Excursion): Fourths and Thirds of Rectangles Students use color tiles to find rectangles that can be divided in fourths and thirds. They copy their rectangles on graph paper and record the total number of tiles and the number of each color used. Then students make fraction flags that show thirds or fourths.

Mathematical Emphasis

- Constructing arrays to represent numbers and identifying halves of the arrays
- Investigating halves of three-dimensional solids
- Constructing two-dimensional arrays that are divided into thirds and fourths
- Describing fractional parts of an array as fractions of a rectangular region
- Describing fractional parts of an array as fractions of the set of tiles used to construct the array
- Designing and constructing a rectangular region that is divided into halves, thirds, or fourths

What to Plan Ahead of Time

Materials

- Geoblocks: 2 sets, each divided into 2 sub-sets as described in Investigation 1, Other Preparation, p. 4 (Sessions 1–2)
- Color tiles: 1 tub per 6 students (Sessions 1–2); 1 tub per pair (Sessions 6–8)
- Crayons or markers to match the colors of the tiles (Sessions 1–2, 6–8)
- Chart paper (Sessions 1–5, 7–8)
- Scissors (Sessions 3–5)
- Overhead projector (Sessions 7–8, optional)
- Construction paper: about 100 sheets (Sessions 3–8)
- Envelopes or resealable plastic bags: 12 (Sessions 3–8)
- Paste or glue sticks (Sessions 3–8)
- Pictures of flags showing halves (Session 6, optional)
- Drawing paper: 2–4 sheets per student (Sessions 6–8)
- Transparent color tiles (Sessions 7–8, optional)

Other Preparation

- Duplicate student sheets and teaching resources (located at the end of this unit) in the following quantities. If you have Student Activity Booklets, copy only the items marked with an asterisk.

For Sessions 1–2

Student Sheet 7, Build the Geoblock (p. 176): 1 per student (optional)

Student Sheet 16, Inch Graph Paper (p. 185): about 4 per student, (class); 2 per student (homework)

Student Sheet 21, Halves and Not Halves (p. 192): 1 per student

Student Sheet 22, Things That Come in Halves (p. 193): 1 per student (homework)

For Sessions 3–5

Shape Halves (p. 196): 2 per student, plus some extras.* Cut out the shapes and store copies of each shape in an envelope or resealable plastic bag. Paste one copy of the shape on the front of each envelope (or bag) so students can see what they are choosing. (You may want to have students or parent volunteers help cut out the shapes.)

Student Sheet 23, Designing Shapes That Can Be Cut in Half (p. 194): 1 per student (homework)

For Session 6

Student Sheet 16, Inch Graph Paper (p. 185): 1–2 per student (class), plus 1 per student (homework)

Student Sheet 24, Half-and-Half Flags (p. 195): 1 per pair of students (homework)

For Sessions 7–8

Student Sheet 16, Inch Graph Paper (p. 185): 1–2 per student

- Cut about 6 paper rectangles, 3" by 4", for a class demonstration. (Sessions 3–5)
- Cut 3" paper squares from plain paper, about 5 per student. (Sessions 3–5)
- Cut 2" or 3" paper squares using red, blue, green, and yellow construction paper (or colors to match the tiles). Plan on 12 squares per student. (Session 6)
- Try to obtain posters or other sources showing flags of many nations. Those showing flags divided in half would be useful. (Session 6)
- Cut 2" or 3" squares of construction paper in different colors. Plan on 8 squares per student. (Sessions 7–8)

Halves of Rectangles and Solids

Materials

- Color tiles
  (1 tub per 6–8 students)
- Prepared sets of
  Geoblocks (1 per group)
- Student Sheet 7
  (1 per student, optional)
- Student Sheet 16
  (about 4 per student,
  class; 2 per student,
  homework)
- Student Sheet 21
  (1 per student)
- Student Sheet 22 (1 per
  student, homework)
- Crayons or markers to
  match the colors of the
  tiles
- Chart paper

What Happens

Students use two colors of tiles to make rectangular arrays that are half one color and half another. They use two colors to color the same rectangle in two ways—to show halves and not halves. They find Geoblocks that are one-half of another Geoblock. Choice Time is used as a Teacher Checkpoint. Their work focuses on:

- constructing two-dimensional arrays that are divided into halves
- constructing arrays to represent numbers and identifying halves of the arrays
- representing rectangles using two colors to show halves and not halves
- investigating halves with three-dimensional solids

Start-Up

Today's Number Sometime during the school day, students express Today's Number using doubles in their number sentences. For example, if the number they are working on is 98, possible combinations include: 49 + 49, or 45 + 45 + 4 + 4. Add a card to the class counting strip and fill in the next number on the blank 200 chart. For complete details on this routine, see p. 116.

Activity

Half-and-Half Rectangles

Distribute color tiles and Student Sheet 16, Inch Graph Paper, to students. Students also need crayons or markers to match the colors of the tiles.

In the last few math sessions you have made many different rectangles. Today we're going to make rectangles that follow a special rule—they will be half one color and half another color.

Make a rectangle with 6 tiles. Make one-half of the rectangle [red] and one-half of it [green]. When you have made your rectangle, draw it on your graph paper and color it to match the tiles.

Observe students as they work. Do they automatically take 3 tiles of each color? Do they seem to make their rectangles somewhat randomly, one tile at a time? What shape rectangle do students make—a 2-by-3 or a 1-by-6? Are they unable to keep track of both aspects of the task? Do they forget to think about the total number of tiles or that half must be each color?

After students have made and recorded their rectangles, call the class together.

Everyone has made a rectangle using 6 tiles that is half one color and half another color. Do you think all the rectangles look the same?

Students can show their drawings to the class. After each student shows his or her rectangle, ask questions such as the following.

■ **Does this rectangle use six tiles? Is one-half of the rectangle [red] and one-half [green]?**

■ **Who has a rectangle like this one? Who has a different rectangle?**

Expect students to disagree about whether a rectangle has been divided into halves. See the **Teacher Note**, Halves of Rectangles (p. 80), and the **Dialogue Box,** What Is a Half? (p. 81), for examples of diverse thinking on what constitutes a half.

Unequal Parts If you have not yet talked about a rectangle where the two parts are *not* equal, make a rectangle such as this one.

Is this rectangle one-half [red] and one-half [green]? Why or why not?

Next, ask students to try making a rectangle that is half one color and half another color using 7 tiles.

Notice how students approach this task. Do some students know, even without trying, that it's impossible? Do other students start manipulating and arranging tiles?

After several minutes, stop students and ask them what they found. If all students agree that it's an impossible task, accept their statement. If some students make a rectangle with 3 tiles of one color and 4 of the other and say that it's about half and half, tell them that for this problem, the number of each color tile must be exactly the same. If some students believe that it really is possible but that they just haven't found out how to do it yet, don't try to convince them otherwise. These students will need more time manipulating the tiles to convince themselves.

Getting Ready for Choice Time   Tell students that during Choice Time they will work with a partner and investigate which numbers they can use to make half-and-half rectangles. Students can work with the numbers they used previously, 8 to 36, using color tiles to build rectangles for the numbers.

If you find a way to make a rectangle half one color and half another color, copy the rectangle on the student sheet and color it. If you find more than one way to make a half-and-half rectangle, show that way on the student sheet also. For each rectangle you draw, write a sentence under your drawing and fill in the blanks.

Write one of the following sentences on the board and read it with the class:

I used ___ tiles. One-half is (color).

One-half is (color). Each half has ___ tiles.

or

I used ___ tiles. $\frac{1}{2}$ is (color). $\frac{1}{2}$ is (color).

Each half has ___ tiles.

You may want to check to see whether students know how to write the notation ½. Although the meaning of the numerator and the denominator is not emphasized in this unit, you might point out that the 2 in the number means that a whole is divided into two equal parts.

Label a sheet of chart paper "Impossible Half-and-Half Tile Rectangles." Post the chart and explain that if students find it impossible to make a half-and-half rectangle for a number, they are to write the number on the chart.

Activity

Introducing Halves of Geoblocks

Gather students in a circle around a set of Geoblocks. Students should be able to see all the blocks in the set. Hold up the square prism that measures 8 cm by 4 cm by 4 cm.

Is there a Geoblock that is one-half of this block?

Let a few students try to find such a block, or hold a 4-cm cube and ask:

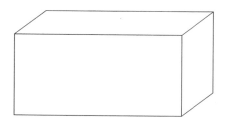

Is this block (cube) one-half of this block (prism)? How could you prove it? Could there be a different block that also is half of this block?

Have a student show that two cubes can be put together to make the prism.

Note: There are three Geoblocks that are half of the 8-cm-by-4-cm-by-4-cm rectangular prism: a 4-cm cube; a rectangular prism, 8 cm by 4 cm by 2 cm; and a wedge, 8 cm by 4 cm by 4 cm.

Students will continue this activity during Choice Time. Tell them that they will look for blocks that are one-half of another block—two identical blocks that when put together will form another Geoblock. When they find blocks that are half of other blocks, they should keep the blocks together until you come over to see what they have done. No more than about six students should share each set of Geoblocks.

Post the following list on the chalkboard:

> 1. Half-and-Half Rectangles
>
> 2. Halves of Geoblocks
>
> 3. Halves and Not Halves

Students will work on these activities for the remainder of Sessions 1 and 2. Make sure students understand Choice Time activities 1 and 2. As they finish one of these activities, introduce some students to Choice 3. You may want to assign and keep track of which students do the Geoblock activity, since no more than about six students should share one (half) set of blocks.

Choice 1: Half-and-Half Rectangles

Materials: Color tiles; Student Sheet 16, Inch Graph Paper; crayons or markers to match the tiles

Students choose a number from 8 to 36 and use two colors of tiles to see whether they can make a rectangle that is half one color and half another. They copy each half-and-half rectangle they find for the number on Student Sheet 16, Inch Graph Paper. After drawing and coloring each rectangle, students write a statement below it to describe the rectangle. See Teacher Note, Halves of Rectangles (p. 80), for examples of what you may see.

If students find a number for which it is impossible to make a half-and-half rectangle, they write the number on a chart labeled "Impossible Half-and-Half Tile Rectangles."

When students have finished making half-and-half rectangles for one number, they choose a different number to investigate, recording each rectangle on graph paper. Some pairs may prefer to find many different half-and-half arrangements for a particular number. Others may prefer to find only one arrangement before investigating a different number.

Choice 2: Halves of Geoblocks

Materials: Geoblocks; Student Sheet 7, Build the Geoblock (optional)

Students find blocks that are one-half of another block. They prove this by putting two identical blocks together and placing them by the Geoblock that is twice as large.

You may need to help students get started on this choice. If you wish more structure for this activity, you can have students use Student Sheet 7, Build the Geoblock (used in Investigation 1). For each block pictured, students look for blocks that are one-half of that block. Alternatively, set out a specific block and ask students to find each block that is half of it. (For many of the blocks, there are several possibilities.)

Choice 3: Halves and Not Halves

Materials: Student Sheet 21, Halves and Not Halves; crayons in two colors; color tiles

Students use crayons in two colors to color a pair of rectangles that have the same dimensions so one shows halves and the other does not. They may use color tiles to help them plan how to color the rectangles.

Activity

Teacher Checkpoint

Halves

Observing the Students

Use this Choice Time as a Teacher Checkpoint to gain information about what students understand about halves. As you observe students working on each activity, use these questions to guide your observations.

- Do students make sense of halves in the different tasks?
- Do students understand that halves are always two parts and the parts are always equal?

Halves of Geoblocks

Consider these questions as you observe students working with Geoblocks.

- Have students found more than one block that is half of another block? Do students recognize that all the half blocks are the same size? You may want to ask, "Do all three of these blocks (that are half of this block) contain the same amount of wood, or does one block have more wood than the others?" This may be a difficult question for many students, so don't expect students to come to agreement about this.

Halves and Not Halves

Consider these questions as students color to show halves and not halves.

- Is this activity easy for students? Do they have to think a lot about how to color the rectangles appropriately?
- Can students justify why they know they have colored in the rectangles in halves or not-halves? If students are having difficulty, ask questions such as, "If I colored 2 squares red and 4 squares green, would it be half red and half green? Why or why not?" Some students may color in the rectangles all the same (or in a similar) way, for example, coloring the left half red and the right half blue. Other students may vary the ways they color in the rectangles, trying to make each rectangle different.

Sessions 1 and 2 Follow-Up

🏠 **Homework**

Half-and-Half Rectangles   After Session 1, students take home copies of Student Sheet 16, Inch Graph Paper, to find more half-and-half rectangles for a number they are investigating or a number that was not investigated during the class session. Students should find at least one rectangle that *can* be divided in half. You may want to have students write "halves" and the number they are investigating at the top of Student Sheet 16 before they take it home.

Things That Come in Halves   After Session 2, students look for examples of things around the house that are in halves. They either bring those examples to school or draw pictures of them on Student Sheet 22, Things That Come in Halves.

As students work on Half-and-Half Rectangles, you are likely to see several different kinds of responses.

■ Some students will make rectangular halves. They may think that only these shapes are "halves."

■ Some students will make congruent, non-rectangular halves.

■ When working with a number of tiles such as 12, some students may make noncongruent halves.

■ Some students will make more than two pieces, explaining that there are the same number of [red] tiles as [green] tiles.

What to accept as the meaning of ½ depends on the context in which the term is being used. To divide a whole into halves, we usually assume that we are dividing something into two equal-size pieces. When we talk about half of a set of objects, we mean that there will be two groups with the same number of objects in each group.

In Sessions 1 and 2, as students work on the activities Half-and-Half Rectangles and Halves and Not Halves, they are dividing both a rectangular array of tiles and a whole (a rectangular region) into halves. Some students will view the problem as dividing a whole into halves, and some will view the problem as partitioning a set of squares into halves. These sessions should help students integrate these two views.

For students who do not accept noncongruent halves or halves consisting of unconnected parts, you might rearrange tiles to see what they will accept.

Is this rectangle half [red] and half [green]? How do you know? How many [red] tiles are there? How many [green]?

How about now? Is this rectangle half [red] and half [green]? How do you know? How many [red] tiles are there? How many [green]?

How about now? Is this rectangle half [red] and half [green]? How do you know? How many [red] tiles are there? How many [green]?

Note that even though you might discuss using the number of squares to judge halves, some students may not be able to accept this criterion. For them, the halves must be visually the same, or the halves must be capable of being physically separated (as if two people were sharing a candy bar). In the latter case, you might discuss how some candy bars are separated into squares and ask if the only way to share is to break it down the middle. You might ask, "What if I gave you three separate squares and me three separate squares (from a candy bar with six squares)? Would we each have one-half of the candy bar?"

What Is a Half?

In this discussion during the activity, Half-and-Half Rectangles (p. 74), students discuss different arrangements for rectangles that are half one color and half another.

Angel: My rectangle has 1 red row and 1 green row. I put 3 tiles in each row.

So each row is a half, and the rows are the same size. Let's look at a different rectangle. Let's look at that long skinny one. Is it a half-and-half rectangle?

Tim: Yes.

How can you tell?

Tim: It has 3 red tiles and then 3 green tiles.

Karina: Mine's just one row, but half of the row is red, and half of the row is green.

Look at Paul's rectangle. How is Paul's rectangle like Angel's?

Paul: It has two rows.

Laura: It has 6 tiles, and it's a rectangle.

Is it the same shape?

Students: Yes.

Is it half and half?

Phoebe: No, you can't make it like that.

Laura: Yes you can.

Paul: It's half and half, but it's in a pattern.

Look carefully at Paul's rectangle and at Angel's rectangle. How many red tiles does each rectangle have? How many green tiles ?

Paul: 3 reds and 3 greens.

Phoebe: They each have 3 red tiles and 3 green tiles.

How are their rectangles different?

Jess: The tiles are checkered in Paul's and in rows in Angel's.

Paul's half and Angel's half have different shapes, but each half has the same number of tiles. Phoebe, if you and Paul shared his rectangle, with you taking the red tiles and Paul the green, would each of you have one-half?

Cutting Congruent Halves

Materials

- Paper rectangles, 3" by 4" (about 6 for class demonstration)
- Construction paper (about 2 sheets per student)
- Prepared Shape Halves
- Paper squares, about 3", cut from plain paper (about 5 per student)
- Student Sheet 23 (1 per student, homework)
- Scissors
- Paste or glue sticks
- Chart paper
- Envelopes or resealable plastic bags

What Happens

Students fold and cut different shapes into congruent halves, then make a shape of their own that can be cut and folded in half. Near the end of Session 5, they discuss which numbers will make half-and-half rectangles. Their work focuses on:

- folding and cutting shapes into congruent halves
- constructing a shape that can be folded and cut into congruent halves

Start-Up

Homework   Students share examples of things they found around the house that are in halves. You may want to start a list or a collection of objects that come in halves.

Quick Images   Using the dot arrays, flash an image on the overhead projector. Ask students to tell how many, and how the arrangement of dots helped them figure out how many were shown. See p. 125 for complete details on this routine.

Activity

Introducing Shape Halves

Introduce the new Choice Time activity.

Mr. Shape-O has a shape shop where he makes shapes. Children often take one shape to share between them. He makes shapes that can be cut in half. The halves are the same size and shape. Some shapes can be cut in half in more than one way, and some can be cut in half in only one way.

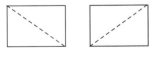

Hold up a paper rectangle and ask a volunteer to fold it in half, then cut it. Ask if anyone can cut it in half a different way. After each cutting, compare to see whether the two halves are the same. If it comes up, discuss whether halves are the same if they are cut in a different direction. For example, are diagonal halves the same if one shape was cut from the top left to the bottom right and another from the bottom left to the top right?

If a student cuts the rectangle into two pieces that are not equal, let another student pick which of the pieces they would prefer and ask if the sharing is fair.

If you wish, introduce or review the word *congruent* by placing one half on top of the other.

These two halves are congruent—they are exactly the same size and shape. When I place one half on top of the other half, they match exactly.

Show students the prepared envelopes containing the shapes cut from Shape Halves.

Mr. Shape-O has made 12 different shapes. Choose one shape that you would like to share and try folding and cutting it into congruent halves. After you have cut a shape in half one way, take another one from the envelope and see if you can find a different way to cut it in half. Glue the halves on a sheet of construction paper. If you have cut a shape in halves in more than one way, show all the ways for the same shape on one sheet of paper.

Students can work with one shape now, finding ways to cut it in half and pasting the halves on a sheet of paper. They can continue working with other shapes during Choice Time.

Activity

Choice Time

Add the new Choice Time activity to the list. For a review of the materials, setup information, and suggestions on what to watch for as you observe students for Half-and-Half Rectangles, Halves of Geoblocks, and Halves and Not Halves, refer to p. 77.

> 1. Half-and-Half Rectangles
>
> 2. Halves of Geoblocks
>
> 3. Halves and Not Halves
>
> 4. Shape Halves

Students will continue Choice Time through about half of Session 5. At that time a discussion will take place on Half-and-Half Rectangles. Monitor students to be sure they have completed this activity by that time.

Choice 4: Shape Halves

Materials: Prepared Shape Halves, scissors, 3" paper squares, construction paper, paste or glue sticks

Students select three or four different shapes from the student sheets and cut them into congruent halves. They see if they can find more than one way to make congruent halves for each shape. They paste their shapes on construction paper, putting all halves for the same shape on the one paper.

When students have finished cutting three or four shapes into congruent halves, they use paper squares to design their own shapes that can be cut into halves. Pose a problem such as the following:

Mr. Shape-O is always looking for new shape designs. What would be a good shape for his shop? Design your own shape. Remember, you must be able to cut your shape into two pieces that are the same size and shape.

Observing the Students

Shape Halves

Use the following questions as a guide while you observe students working.

■ Is it obvious to students how to cut the shapes in halves? After cutting a shape in half one way, do they look for other ways to cut it?

Observe how students design their own shapes.

■ What kinds of shapes do they draw? Ask them how they know their shapes can be cut into congruent halves.
■ Do students first draw the entire shape and then figure out how to cut it? Do any students fold a paper in half and then cut out a shape? (They will be doing a similar activity in the next investigation.)

Students can display their shape designs for others to see. You may want to trace the shapes on paper and duplicate them. Then students can cut one another's shapes into congruent halves.

Activity

Class Discussion: Which Rectangles Make Halves?

About 15 minutes before the end of Session 5, ask students to take out all the half-and-half rectangles they made.

There are a couple of ways you can have students share their results. You may wish to do some combination of the following.

- Review each number in sequential order to see which numbers students were able to build half-and-half rectangles for and which they found impossible. Record the numbers in two lists on the board. Also keep track of the numbers that no one investigated. (These could be homework.)

- Ask students who found half-and-half rectangles for the same number to get together to compare their rectangles and see if they (1) made the same rectangle, and (2) colored them in the same half-and-half pattern. For example, some students investigating the number 12 may have made a 3-by-4 rectangle and other students may have made a 2-by-6 rectangle. Even if students made the same rectangle, there are many ways to make them half one color and half another.

Ask students to look for a pattern that helps them predict which numbers can be made into half-and-half rectangles and which numbers cannot. On chart paper, start a two-column class chart labeled *Halves* and *Not Halves*. If you want to list numbers sequentially, ask students to make half-and-half rectangles for the numbers 1, 2, 3, 4, and 5, and review what they found out for 6 and 7, when the activity was introduced.

Halves	Not Halves
2	1
4	3
6	5
8	7
10	9

Sessions 3, 4, and 5 Follow Up

Activity

Designing Shapes That Can Be Cut in Half   After Session 4, students design shapes that can be cut in half. They draw pictures of the shapes on Student Sheet 23, Designing Shapes That Can Be Cut in Half.

 **Homework**

Fraction Flags

Materials

- Pictures of flags showing halves (optional)
- Color tiles (1 tub per pair)
- Student Sheet 16 (1–2 per student, class; 1 per student, homework)
- Drawing paper (1–2 sheets per student)
- Crayons or markers
- Construction paper squares 2" or 3" (12 per student)
- Paste or glue sticks
- Student Sheet 24 (1 per student, homework)

What Happens

Students make rectangular flags that are divided into halves with each fractional part of the flag a different color. They decide on the size of their flag and how to divide it in half. Students use color tiles to design their flags, then copy the design by gluing construction paper squares on large paper. Their work focuses on:

- constructing a rectangular region that is divided into halves
- copying and enlarging a design

Start-Up

Quick Images   Use the 10 Frames images and show *two* numbers on the overhead projector at the same time. Students write down the sum of the two and/or the addition expression the numbers represent. For complete details, see p. 125.

Today's Number   Sometime during the school day, students brainstorm ways to express Today's Number. Encourage students to use both addition and subtraction in their number sentences. For example, if Today's Number is 100, possible expressions include: 60 + 60 – 20 or 90 – 40 + 50. Add a card to the class counting strip and fill in the next number on the blank 200 chart. For complete details, see p. 116.

Activity

Fraction Flags

If you have pictures of flags that are divided in halves, begin this activity by showing these to students. (Posters showing national flags are commercially available. Most almanacs and encyclopedias contain color inserts of flags.)

These are pictures of flags from different countries in the world. [*Point out the flags and name the countries they represent.*] **Some of these flags remind me of our work with rectangles.** [*Point out a few flags divided in halves.*] **What shape are these flags? How are they like some of the rectangles you have been working with? Do you see other flags divided in halves?**

Students may volunteer that flags and their rectangles have the same shape, but most students may not be familiar with different kinds of flags and may not make the connection that both flags and their rectangles can have halves.

Today we are going to design and make flags. Your flag will be a rectangle that has halves. Use the color tiles and graph paper to design your flag. Then make your flag by gluing these larger paper squares onto drawing paper. Use two colors and make half your flag each color.

Have students first plan and design their flags using color tiles and graph paper. The color tiles can be arranged in a pattern or design, but each color should represent one-half of the design. Students can record their flag by coloring spaces on the graph paper.

As students design their flags, ask them to describe their flags to you. Remind them to make sure their flag designs show halves before they make their flags with the paper squares.

You might display students' completed flags in a school corridor, or hang them from the classroom ceiling (by attaching them to a string going across the room). As students compare the flags, ask questions such as the following.

- What sizes are the rectangles that have been made into flags?
- How are the flags designed so they are half one color and half another?
- How are flags that are the same size alike and how are they different?
- How are flags that are different sizes similar?

Session 6 Follow-Up

🏠 **Homework**

Half-and-Half Flags   Students use Student Sheet 16, Inch Graph Paper, to design a fraction flag at home. The flag can include a pattern or design, but each color should represent one-half. Students also fill in the blanks on Student Sheet 24, Half-and-Half Flags, with information about their flags.

▨ **Extensions**

Congruent-Halves Flags   Students make a second rectangular flag that has two congruent halves using construction paper squares, crayons, or paint. Students plan their flags first on smaller paper.

Halves Display   Students make a display of things in halves and write on index cards to identify them. Some possibilities are a jar half full, a bag of blocks half of one color and the other half another color, a half-eaten cookie, a mark halfway across the top of a table and another mark halfway up the leg of the table.

Fourths and Thirds of Rectangles

What Happens

Students use color tiles to find rectangles that can be divided in fourths and thirds. They copy their rectangles on graph paper and record the total number of tiles and the number of each color used. Then students make fraction flags that show thirds or fourths. Their work focuses on:

- constructing two-dimensional arrays that are divided into thirds and fourths
- describing fractional parts of an array as fractions of a rectangular region
- describing fractional parts of an array as fractions of the set of tiles used to construct the array

Materials

- Color tiles
  (1 tub per pair)
- Student Sheet 16
  (1 or 2 per student)
- Crayons or markers
- Construction paper squares, 2" or 3" in several colors
  (8 per student)
- Drawing paper
  (1–2 sheets per student)
- Overhead projector
  (optional)
- Transparent color tiles
  (optional)
- Chart paper
- Paste or glue sticks

Activity

Fourths and Thirds of Rectangles

Distribute a container of color tiles to each pair of students.

We have been making rectangles that are in halves—half one color and half another color. Today we're going to make rectangles that are in fourths. Who can tell us what *fourths* means?

Students may interpret *fourths* as four parts or four colors. Question students to see if they understand that the parts have to be equal or that they will use the same number of tiles of each color.

You will make rectangles using four colors of tiles. Use the same number of each color tile. Begin by using eight tiles, making a rectangle that has four colors. Each color should be one-fourth of the rectangle.

Observe students as they work and evaluate their understanding. Is this task easy and obvious for students? Do they make a rectangle of eight tiles that is in halves (four tiles of one color and four of another color), then try to work in the other two colors? Do they take eight tiles each of four colors and try to make a rectangle?

As you talk, emphasize *fourths* rather than eight tiles.

Remember, you need to make a rectangle that is in fourths—four colors, with the same number of tiles of each color.

When students have finished, ask volunteers to share their strategies with the class. Use students' ideas to promote thinking, even if the outcome is not what you expected. For example, if a student reports using eight red tiles, eight green tiles, eight blue tiles, and eight yellow tiles—not the task you assigned, but a rectangle divided into fourths—you might respond:

Did Tim make a rectangle that is in fourths? [*Tim can make his rectangle on the overhead if you have transparent color tiles, or have students come and look at his rectangle.*] **Instead of using eight tiles, how many tiles did Tim use in his rectangle that is one-fourth red, one-fourth green, one-fourth blue, and one-fourth yellow?**

After students share how they made rectangles using eight tiles, write the following on the board. Students can complete what goes in the blank.

We used ___ tiles. $\frac{1}{4}$ is (color), $\frac{1}{4}$ is (color), $\frac{1}{4}$ is (color), $\frac{1}{4}$ is (color).

Each fourth has ___ tiles.

Ask students what they think the 4 tells you in the ¼ notation.

You might want to ask students if they can make a rectangle in fourths using 10 tiles. After a few minutes, acknowledge that, just as for halves, there are some rectangles that are impossible to show in fourths.

Next, talk about rectangles that show thirds.

Some rectangles can be made to show thirds—one-third of the rectangle will be a different color tile. What do you think *thirds* means?

Make a rectangle using three colors of tiles. Use the same number of each color tile.

Let students decide how many tiles to use in their rectangle. If they need suggestions on how to begin, tell them to make a rectangle using 6 tiles.

Again, ask volunteers to share their strategies. When a student has made a rectangle in thirds, record the results on the board.

<u>(Name)</u> used ___ tiles. $\frac{1}{3}$ is <u>(color)</u>, $\frac{1}{3}$ is <u>(color)</u>, $\frac{1}{3}$ is <u>(color)</u>.

Each third has ___ tiles.

Display two sheets of chart paper that have been labeled *Fourths* and *Thirds.* You can structure this activity in several ways. Suggest the following for an open investigation.

Use color tiles to make rectangles that have thirds or fourths. Rectangles in thirds use three colors of tiles with the same number of each color. Rectangles in fourths use all four colors of tiles with the same number of each color. When you have found a rectangle that is thirds or fourths, copy it on Student Sheet 16, Inch Graph Paper, and write a description under it.

I used ___ tiles. $\frac{1}{3}$ is <u>(color)</u>, $\frac{1}{3}$ is <u>(color)</u>, $\frac{1}{3}$ is <u>(color)</u>. Each third has ___ tiles.

I used ___ tiles. $\frac{1}{4}$ is <u>(color)</u>, $\frac{1}{4}$ is <u>(color)</u>, $\frac{1}{4}$ is <u>(color)</u>, $\frac{1}{4}$ is (color). Each fourth has ___ tiles.

The number of options can be limited further by asking students to look specifically for thirds or fourths or by limiting the number of tiles they use to make rectangles.

Observing the Students There may be a great deal of diversity among your students. If some do not understand the thirds and fourths activity, encourage them to continue to work with halves. It may help these students to work with partners.

Challenge other students to think more about the relationships among the rectangles they make by asking questions such as the following.

- **Are there some numbers that you can make into rectangles to show halves, thirds, *and* fourths? Which ones?**
- **Are there some numbers that can't be made into rectangles to show halves, thirds, or fourths? Which ones?**
- **What number patterns do you see when you make rectangles that show thirds?**
- **What number patterns do you see when you make rectangles that show fourths?**

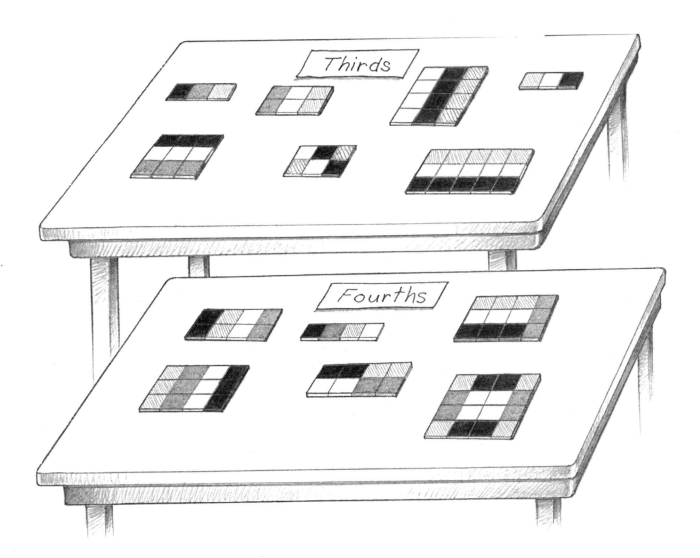

Students make rectangular flags that are divided into thirds or fourths using construction paper squares. As for the halves flags, they decide on the size of their flags and use three or four colors of paper squares to make their flags.

Thirds and Fourths Flags

You can use three colors and make one-third of your flag in each color, or you can use four colors and make one-fourth of your flag in each color. First design your flag using color tiles. Then make your flag, just like you did with your halves flags.

When students have made their flags, they can write short reports describing their flags and why they know their flags are thirds or fourths.

Sessions 7 and 8 Follow-Up

Thirds and Fourths Displays   Students make a thirds display and a separate fourths display. These displays may include items divided and labeled $1/3$ to $2/3$ and $1/4$ to $3/4$, as well as some things divided into three $1/3$'s and others into four $1/4$'s.

Extension

INVESTIGATION 4

Symmetry

What Happens

Sessions 1 and 2: Symmetrical Designs Students observe the building of a symmetrical pattern block design, either on or off the computer, and discuss its characteristics. They look for symmetrical things in the world. They build symmetrical designs using pattern blocks, Geoblocks, and the *Shapes* software.

Sessions 3 and 4: Reflecting Blocks and Tiles Students continue to make symmetrical designs using mirrors, pattern blocks, color tiles, and the *Shapes* software.

Sessions 5 and 6: Paper Folding and Cutting Students cut symmetrical shapes from folded paper and work on Choice Time activities. Near the end of Session 6, students share the symmetrical designs they made with different materials. They use the fold-and-pinch procedure to check whether their designs are symmetrical.

Session 7: Symmetrical Pictures Students make pictures using symmetrical shapes cut from folded paper. They write an explanation of how they know their shapes are symmetrical and define line symmetry.

Mathematical Emphasis

- Finding and describing objects that have mirror symmetry
- Making two-dimensional symmetrical designs
- Building three-dimensional symmetrical structures

What to Plan Ahead of Time

Materials

- Pattern block stickers (Sessions 1–4, optional)

- Transparent pattern blocks (Sessions 1–4, optional)

- Pattern blocks: 1 tub per 6–8 students (Sessions 1–6)

- Geoblocks: 2 sets, each set divided into 2 subsets as described in Investigation 1, Other Preparation, p. 4 (Sessions 1–6)

- Computers: Macintosh II or above, with 4 MB of internal memory (RAM) and Apple System Software 7.0 or later: 1 per 4–6 students (Sessions 1–6, optional)

- Apple Macintosh disk, *Shapes* (Sessions 1–6, optional)

- Projection device or large-screen monitor on one computer for whole-class viewing (Sessions 1–2, optional)

- Crayons or markers (Sessions 1–4, 7)

- Overhead projector (Sessions 1–4, 7, optional)

- Mirrors: 1 per 3–4 students (Sessions 3–4)

- Scissors (Sessions 5–7)

- Plain paper (Sessions 3–4)

- Color tiles: 1 tub per 6–8 students (Sessions 3–7)

- Construction paper: about 4 sheets per student (Sessions 5–7)

- Letter-size paper: 3 half sheets per student (Sessions 5–7)

- Paste or glue sticks (Sessions 5–7)

- Chart paper (Session 7, optional)

Other Preparation

- Duplicate student sheets and teaching resources (located at the end of this unit) in the following quantities. If you have Student Activity Booklets, copy only the transparency marked with an asterisk.

  For Sessions 1–2

  Student Sheet 25, Looking for Symmetry (p. 198): 1 per student (homework)

  For Sessions 3–4

  Student Sheet 16, Inch Graph Paper, (p. 185): 5 per student, plus a transparency*

  Student Sheet 26, Exploring Mirror Symmetry (p. 199): 1 per student (homework)

  For Session 7

  Student Sheet 27, Fold and Cut (p. 200): 1 per student (homework)

- If you are using computers with this unit, try the Mirrors activity on the *Shapes* software before demonstrating it for students. See the Teacher Tutorial (p. 129) for additional information on the software. The **Teacher Note**, Introducing the *Shapes* Software (p. 36), provides suggestions on introducing the software to your class. (Sessions 1–2)

- Cut enough letter-size paper in half so you have 3 half sheets per student. You will also need 3 half sheets per student for Session 7, so you may want to cut enough for all sessions at this time. (Sessions 5–6)

- If you have not already done so, cut enough letter-size paper in half so you have 3 half sheets per student. (Session 7)

Symmetrical Designs

What Happens

Students observe the building of a symmetrical pattern block design, either on or off the computer, and discuss its characteristics. They look for symmetrical things in the world. They build symmetrical designs using pattern blocks, Geoblocks, and the *Shapes* software. Their work focuses on:

- describing objects and patterns that have mirror symmetry
- identifying objects that have mirror symmetry
- making symmetrical patterns and designs
- building three-dimensional symmetrical structures

Start-Up

Today's Number Suggest that students use addition and subtraction in their number sentences. For example, if Today's Number is 102, possible expressions include: $50 + 60 - 8$ or $90 - 10 + 22$. Add a card to the class counting strip and fill in the next number on the blank 200 chart. See p. 116 for complete details on this routine.

Materials

- Computers with *Shapes* installed (optional)
- Pattern blocks (1 tub per 6–8 students)
- Prepared set of Geoblocks (1 set per group)
- Pattern block stickers (optional)
- Crayons or markers
- Transparent pattern blocks (optional)
- Overhead projector (optional)
- Student Sheet 25 (1 per student, homework)

Activity

Symmetry in the World

This activity, which introduces the idea of symmetry, can be done either on or off the computer. If you are using computers with this unit, introduce both of the following activities to students as they will have the opportunity to work on each one during Choice Time.

Introducing Symmetry on the Computer Use the Mirrors activity on the *Shapes* disk and a projection device for screen display (if available) to introduce the idea of symmetry. Open the Mirrors activity and drag several shapes from the *Shapes* window, one at a time.

When I place a shape on the screen on one side of the line, a shape appears on the other side of the line. What do you notice about the shapes that appear? [*Move a shape to a different position.*] **When I move a shape, what happens to the same shape on the other side of the line?**

Repeat this process several times with different shapes. Then click on one of the Turn tools and turn one of the shapes.

When I turn a shape, what happens to the same shape on the other side of the line?

Students may notice that the other shape turns in the opposite direction. Build a symmetric design, such as the one shown here.

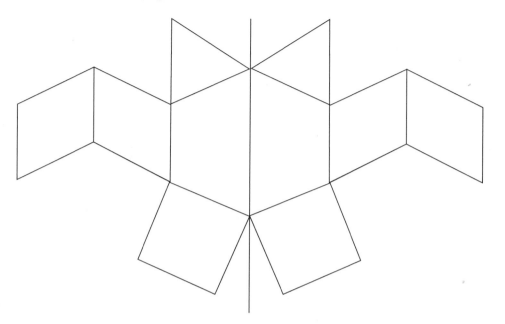

Look at this design. What can you tell me about it?

Students may notice that the design has two of each block—for every pattern block on one side of the center line, there is a corresponding block on the other side. Students may point out that you built one-half of the design—the computer automatically built the other half. Some students may say that the design is balanced or that it's the same on both sides. Introduce the term *symmetry* but encourage students to use their own descriptive language. The **Dialogue Box,** Discussing Symmetry (p. 103), illustrates how students described a symmetric design in one second grade class.

Discuss other things that are symmetrical.

Lots of things in the world are symmetrical. Look around. What are some that you see?

Students share things they see. If they can't think of any, suggest people's or animals' bodies, furniture, or other things in the room. You might ask a student to come to the front of the room and others to describe how a person is symmetrical.

Sometimes it's difficult for students to see or imagine lines of symmetry. For example, they might say things such as, "That bookcase would be symmetrical if we put another one right beside it."

Tell students that in this investigation, the activities will be about symmetry. They will make symmetrical designs with pattern blocks. Students will have the opportunity to work on this computer activity during Choice Time.

Introducing Symmetry off the Computer   The same activity on symmetry can be done with pattern blocks.

Gather students in a circle (or use the overhead and transparent pattern blocks) and make a simple symmetrical design with pattern blocks. Ask students where you should place the blocks to keep the design symmetrical. Your design should have line or mirror symmetry rather than rotational symmetry. For more information about symmetry, see the **Teacher Note,** Making Symmetrical Designs (p. 102).

Continue this activity with the discussion of symmetry found in objects all around, as described above under Introducing Symmetry on the Computer.

Tell students that they will build symmetrical designs with pattern blocks during Choice Time.

Activity

Introducing Geoblock Buildings

Quickly introduce this activity to the whole group or to small groups.

Many buildings are symmetrical. They have parts that match. Let's build a symmetrical building with Geoblocks. [*Start building a simple symmetrical building, such as the one below.*]

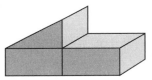

What block should I add to make it symmetrical? Which direction should the block be placed?

Have a student place the block on the building. You might have the student place it so it first slopes one way and then the other. Ask others to tell which way makes the building symmetrical.

Tell students that during Choice Time, one of the activities will be to build symmetrical buildings with Geoblocks.

Post the following list on the chalkboard.

> 1. Mirrors Activity (computer)
>
> 2. Pattern Block Symmetry
>
> 3. Geoblock Buildings

Each of these activities involves building symmetrical patterns. Since only a small number of students can use the Geoblocks at a time, you may want to monitor which students work on Geoblock Buildings. Students will have the opportunity to continue all these activities through Session 6.

1. Mirrors Activity

Materials: Computers with the *Shapes* software installed

Students use the Mirrors activity in the *Shapes* software to build symmetrical designs similar to the one created at the beginning of the session. They drag shapes onto the Work window and use the arrow to slide and the Turn tool and Flip tool to position the shapes as desired. If students like their designs, they can save and print them.

2. Pattern Block Symmetry

Materials: Pattern blocks, crayons or markers, pattern block stickers (optional)

Students work individually or with partners to build a symmetrical design with pattern blocks. When finished, they show their design to someone else in the group and see if all agree that it is symmetrical. If students like their design, they can record it using pattern block stickers.

3. Geoblock Buildings

Materials: Geoblocks

Students build symmetrical buildings with Geoblocks. When they finish, students share their buildings with you and tell you how they know the buildings are symmetrical.

Since the number of students who can use the Geoblocks at one time is limited, you may wish to schedule students' time at this activity throughout this investigation so all have an opportunity to build Geoblock structures.

Observing the Students

As you observe students at work, consider the following questions and suggestions.

Mirrors Activity

Most students will want to explore what happens when they drag shapes and use the motion tools before trying to build a symmetrical design. Encourage them to do so. Suggest they explore all possibilities of the software including other tools, such as the Glue tool. Ask students to describe what happens as they use a tool.

Notice how students build designs.

- Do they randomly place shapes, or do they have a plan?
- Can they explain how they used the tools to build the design?
- Do they know how the design will change if they move a shape?

When students have explored building symmetrical designs using the vertical mirror (that comes on automatically), they can then choose from the **Options** menu to change to the horizontal mirror, or they can use both mirrors and create a four-part design.

Pattern Block Symmetry

- Are students' designs symmetrical?
- Are students able to identify and change blocks that are placed incorrectly?
- Are students' designs simple or elaborate? Do any have more than one line of symmetry? If so, can students identify this?

Suggest to some students that they try to build larger designs that are symmetrical. A challenge such as, "Can you cover your desk or a table with a symmetrical design?" can often turn into a cooperative venture among many students.

Geoblock Buildings

As students are building, ask them to explain how they know their buildings are symmetrical. If some students have difficulty understanding how to build a symmetrical building, build a building together with a student, with the student placing one block at a time while you add corresponding blocks.

Some students may build Geoblock buildings that are symmetrical on four sides. If they do so, encourage them to continue.

Save 10 minutes at the end of some of the next four sessions for students to share with the class what they built with the Geoblocks. Other students can observe the buildings and discuss why the buildings are or are not symmetrical. Consider photographing the students' buildings to create a permanent record of their work that can be posted and/or kept in their math folders.

Sessions 1 and 2 Follow-Up

Looking for Symmetry   Ask students to look around their homes or in magazines and find things that are symmetrical. Start a bulletin board for students to post objects or pictures they find or drawings they make of symmetrical things. Student Sheet 25, Looking for Symmetry, includes directions as well as a plan for students to draw their symmetrical objects.

⌂ Homework

Teacher Note ▷ *Making Symmetrical Designs*

This investigation builds on students' intuitive recognition of symmetry. Students are aware of symmetry in butterflies, hearts, designs, and pictures. They seem to know when something is in balance. They frequently draw or make designs or pictures that are symmetrical.

Designs or patterns that can be flipped or folded over along a line so that one half is on top of the other half have *mirror symmetry*. The two halves are reflections of each other. (Some shapes have more than one line of symmetry or folding lines.)

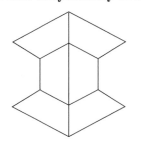

Designs that revolve around a central point have *rotational symmetry* or *circular symmetry*. As a design is rotated, at some point(s) it will fit upon itself before it is completely rotated.

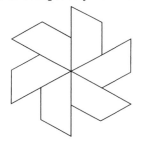

Some designs have both mirror and rotational symmetry.

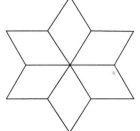

Many shapes are symmetrical. Shapes such as a circle, a regular hexagon like the yellow pattern block, and any even-sided regular polygon have both mirror and rotational symmetry. Other shapes, such as the trapezoid or a heart, have

only mirror symmetry. In Investigation 3, all of Mr. Shape-O's shapes could be cut into congruent halves, but the two halves did not always have both kinds of symmetry.

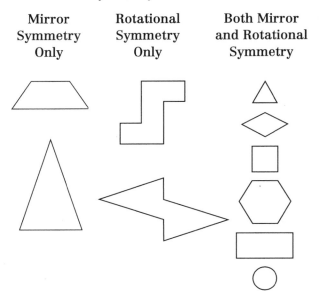

| Mirror Symmetry Only | Rotational Symmetry Only | Both Mirror and Rotational Symmetry |

In this unit we focus on mirror or reflective symmetry. (Students investigate rotational symmetry in *Mathematical Thinking at Grade 4*.) You may observe, however, that some students will make pattern block designs that have rotational symmetry and not mirror symmetry. It is common for students to start pattern block designs with a central hexagon with the design radiating out like a sunburst. Some of these designs might also have mirror symmetry, while some may not.

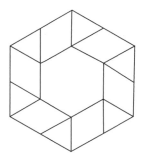

Ask students to describe how the designs do or don't have mirror symmetry. You can point out that some designs have another kind of symmetry, even though you can't fold or flip them to show mirror symmetry.

◼ D ◼ I ◼ A ◼ L ◼ O ◼ G ◼ U ◼ E ☐ B ◼ O ◼ X ◼

Discussing Symmetry

With the class gathered in the meeting area, the teacher introduces symmetry off the computer through the activity Symmetry in the World (p. 96). The teacher places pattern blocks one by one on the floor in front of her, building a symmetric design that has one line of symmetry. She continues to place blocks one by one until there are eight blocks in her design. In the discussion that follows, students share their knowledge of symmetry and use their own language to describe it.

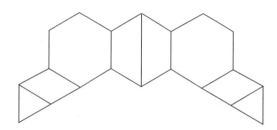

What do you notice about this design? Trini?

Trini: It's symmetric.

Say more about that.

Trini: It's the same on both sides. If you cut it, it's the same on both sides. [*She points out where the two trapezoids meet, an imaginary vertical line of symmetry.*]

Karina: If you add a green triangle and a blue block to each side, then it's the same this way too. [*She shows how to place two more rhombuses and two more triangles so that the design has a horizontal line of symmetry as well as a vertical line.*]

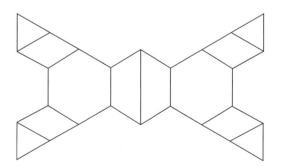

Franco: I know another way too. You can trade the two red trapezoids for a hexagon. Like this. [*He draws an imaginary vertical line down the middle of the hexagon with his finger.*]

Franco's saying that if I put these blocks here, it still looks the same on both sides, but you'd have to cut the block. [*She replaces the trapezoids with a hexagon and then goes back to the original.*]

Let's look at this design. What else can you say about symmetry?

Naomi: It's symmetrical with each side like a mirror.

Jeffrey: First you put a trapezoid down, then another trapezoid, and then the hexagons. If you just put a bunch of shapes down it wouldn't work. You have to plan it.

Laura: If the two sides were twins and this was the middle [*she points to the line of symmetry*], the twins could look through and see themselves.

Those are good ways to explain it, ways I hadn't thought of.

Reflecting Blocks and Tiles

Materials

- Choice Time materials from previous sessions
- Transparent pattern blocks (optional)
- Pattern blocks (1 tub per 6–8 students)
- Mirrors (1 per 3–4 students)
- Color tiles (1 tub per 6–8 students)
- Student Sheet 16 (5 per student)
- Crayons or markers
- Pattern block stickers (optional)
- Overhead projector (optional)
- Transparency of Student Sheet 16 (optional)
- Student Sheet 26 (1 per student, homework)
- Plain paper (1 sheet per pair)

What Happens

Students continue to make symmetrical designs using mirrors, pattern blocks, color tiles, and the *Shapes* software. Their work focuses on:

- describing objects and patterns that have mirror symmetry
- identifying objects that have mirror symmetry
- making symmetrical patterns and designs
- building three-dimensional symmetrical structures

Start-Up

Homework Ask students to share examples of things with symmetry they found at home. If they have drawn or cut out pictures, post them on a bulletin board.

Quick Images Using the 10 Frames images, show two numbers on the overhead at the same time. Ask students mentally to figure out the sum. They can share the visual clues that help them know the sum. For example, a student might say for 5 and 7, "I knew it was 12 because the top row of each was filled so that makes 10, and there were 2 extra." See p. 125 for complete details.

Activity

Mirror Designs

In this demonstration, you will show students how to use the materials for a new activity. Students will work on these activities during Choice Time.

We will use mirrors to make symmetrical designs with pattern blocks. But before we use a mirror with pattern blocks, let's explore using the mirror on other things. A mirror can make interesting designs with ordinary things.

Demonstrate how to stand a mirror perpendicular to a surface such as a picture or book cover and look in the mirror. Show students how to move the mirror around the surface it's on to change the designs they see. For example, you might tell students they can "bend" a line of written words or "turn it upside" down using a mirror.

Then show students how they can use mirrors with pattern blocks. Do this in small groups or distribute several mirrors so students can gather around one of them to observe what happens.

Fold and then open a piece of paper. The fold can be anyplace on the paper—it doesn't have to be horizontal or vertical. Place a pattern block on one side of the fold. Stand a mirror on the folded line, facing the block. Look at the mirror and see the pattern block on the other side of the fold. Remove the mirror and place another pattern block on the other side of the fold, where the reflection was. Check the placement by replacing the mirror.

Move the pattern block to a different location and/or orientation. Ask students to predict where it will be in the mirror by placing a pattern block on the other side of the fold. Check with the mirror.

Students should explore with the mirror and one block first. After they become familiar with reflecting one block in a mirror, suggest that they build a design using three or four pattern blocks on one side.

Demonstrate by having a student make a design. Point out that there are a lot of different patterns they could make, depending on where they put the mirror.

Move the mirror around on the paper until you see a design you like. Then draw a line with a pencil along the edge of the mirror.

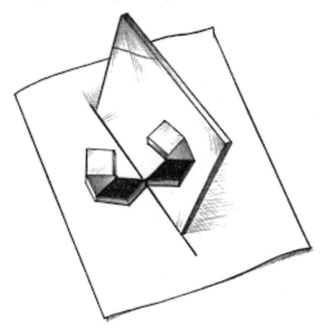

Demonstrate the process. After drawing the line, remove the mirror and build the reflection of the design on the other side of the line drawn on the paper. When you are finished, stand up the mirror again along the line and check to see if the design is complete.

Tell students that they will have the opportunity to work on this activity during Choice Time.

Activity

Introducing Copy Tiles

Briefly introduce Copy Tiles to the class. On an overhead projector, place a transparency of Student Sheet 16 and some transparent pattern blocks. (Or gather students in a circle for the demonstration.) On the student sheet, darken one horizontal or vertical line with a marker. Place one tile in a square along the line and ask a volunteer to "copy" by placing the same-color tile in a corresponding (reflected) square on the other side of the line. Continue playing several rounds.

Tell students that they will have the opportunity to play this game with a partner during Choice Time.

Activity

Choice Time

Add two new activities, Mirror Designs and Copy Tiles, to the activity list. Students will have the remainder of this session and most of the next two sessions to work on these activities.

> 1. Mirrors Activity (computer)
>
> 2. Pattern Block Symmetry
>
> 3. Geoblock Buildings
>
> 4. Mirror Designs
>
> 5. Copy Tiles

For information on materials, setup, and what you might watch for as you observe students working, see the descriptions for Mirrors Activity, Pattern Block Symmetry, and Geoblock Buildings on p. 99.

Choice 4: Mirror Designs

Materials: Pattern blocks, mirrors, pattern block stickers (optional), paper

Students explore what happens with pattern blocks and mirrors. They fold a paper, place a pattern block on one side of the fold, and predict the exact location of the reflected block prior to standing a mirror on the fold line. Then they check their prediction using a mirror.

Next, students build a design with three to five blocks. They explore design variations by moving the mirror around their pattern block arrangement. When they find a design they like, they draw a line on the paper along the edge of the mirror and build the reflected side using more pattern blocks. They then check their building by replacing the mirror on the line. Students make a copy of one design by tracing around their pattern blocks and coloring, or using pattern block stickers.

For a challenge, tell students to build a design, place the mirror, and build the symmetrical half. This design can now be considered a new, larger design, and students again place the mirror beside the design and build another half. Students should first predict how many blocks will be in their new design before they begin building. (Each time, the number of blocks doubles.)

Choice 5: Copy Tiles

Materials: Color tiles; Student Sheet 16, Inch Graph Paper; crayons or markers

Pairs of students fold graph paper along a line, then build a symmetrical design with color tiles. One student places a tile on one side of the line. A partner "copies" by placing the same-color tile in the corresponding (reflected) square on the other side of the fold line. Players can switch roles so each has a turn being the leader and the copier. Students then choose one or two of their designs and record them.

For a variation, each person can make a complete design with several color tiles on one side of a page and color in the squares with crayons or markers. Partners exchange papers and complete each other's designs by building the reflection with color tiles. Collect these papers and start a class set of Copy Tile Task Cards for students to choose and do during Choice Time.

Observing the Students

Observe students as they work. You may want to consider the following questions and suggestions as you observe students.

Mirror Designs

Encourage students to experiment placing and moving a pattern block on a paper until they can predict what will happen to its mirror image. Ask questions such as:

- **Where will the pattern block in the mirror be if you move the block over here?**
- **What will happen to the pattern block in the mirror if you turn the block around this way?**
- **What would happen if you moved the mirror to this location?**

Ask students to describe how their designs change depending on where they place the mirror. Observe how they build the second half of their designs. Do students need to keep the mirror in place to know where to place the blocks?

Copy Tiles

Notice how students work together on this task. Do they take turns making and copying the pattern? How do they resolve disagreements about where a "copied" tile should be placed? How do they talk about their designs to convince others that each side is a mirror image?

Students should save their recorded tile designs for use in Session 5.

Near the End of the Session   Five or 10 minutes before the end of each session, tell students to complete what they are working on, including any necessary recording. Remind students to put their papers in their folders and complete their logs, if they haven't already done so.

Sessions 3 and 4 Follow-Up

Homework

Exploring Mirror Symmetry   Students use a hand mirror at home to explore mirror symmetry. Suggest that they use all sorts of objects they find at home. Students can write a brief description of the most interesting object they found on Student Sheet 26, Exploring Mirror Symmetry.

Remind the students to place the mirror along the midline of the object they think is symmetrical, instead of placing the object next to the mirror, which would give them a mirror reflection of the entire object.

Paper Folding and Cutting

What Happens

Students cut symmetrical shapes from folded paper and work on Choice Time activities. Near the end of Session 6, students share the symmetrical designs they made with different materials. They use the fold-and-pinch procedure to check whether their designs are symmetrical. Their work focuses on:

■ exploring symmetry by folding and cutting paper patterns

■ making symmetrical patterns and designs

■ building three-dimensional symmetrical structures

Materials

■ Letter-size paper, half sheets (3 per student)

■ Construction paper (about 2 sheets per student)

■ Scissors

■ Paste or glue sticks

■ Choice Time materials from previous sessions

Start-Up

Homework Ask students to briefly share their experiences exploring symmetry at home with hand mirrors. Students can share any descriptions they wrote.

Today's Number Suggest that students use three numbers in their number sentences. For example, if the number they are working on is 104, possible combinations include: 50 + 50 + 4 or 40 + 50 + 14. Add a card to the class counting strip and fill in the next number on the blank 200 chart. See p. 116 for complete details.

See p. 116 for complete details.

Activity

Introduce this activity to the whole group. Fold a sheet of paper in half and draw half a pine tree, turtle, or other design on it.

Introducing Fold and Cut

I am going to cut along this outline. What will I have when the paper is unfolded?

Students will probably recognize that you will have a tree or turtle. Cut out the shape, unfold it, and show students.

The tree (turtle) is symmetrical—the folding line (or line of symmetry) divides the tree (turtle) into two halves that are alike. In a new Choice Time activity, you will be cutting symmetrical objects from paper.

Give students an opportunity to describe what kinds of paper objects they have made in the past. If objects students describe are good examples, list them on the board as suggestions of what they can cut from paper. (Some possibilities include: heart, diamond, rectangle, square, tree.)

I just folded a paper and cut out a symmetrical tree. Many of you have also done paper-folding activities. Olga has folded origami paper to make paper boxes and balloons. Imani said he likes to fold paper to make airplanes. Several of you have made kites with large pieces of paper. In this activity, you will try to cut different shapes out of folded paper.

Some students may recognize that this activity is similar to the Mr. Shape-O activity where students cut paper shapes in halves. Acknowledge that in both activities, you have two congruent halves.

Add the new activity to the Choice Time list. Check with students about which activities they have completed and which they plan to do today. Many students may want to work on Fold and Cut. If you are scheduling students through the Geoblocks choice, let the class know whose turn is today. Refer to p. 99 and p. 107 for materials and setup information for activities 1–5.

1. Mirrors Activity (computer)

2. Pattern Block Symmetry

3. Geoblock Buildings

4. Mirror Designs

5. Copy Tiles

6. Fold and Cut

Choice 6: Fold and Cut

Materials: Half sheets of paper, scissors, construction paper, paste or glue sticks

Write the following list of shapes on the chalkboard. Include suggestions made by students during the introduction.

heart	rectangle	square
diamond	triangle	tree
a shape of your own	trapezoid	

Students fold their paper in half, draw, then cut out their shape. Students may make several tries if they are not satisfied with their first attempts. When students have cut a shape they like, they paste the shape and the paper from which it was cut on construction paper.

Observing the Students

Observe students while they work. See p. 100 and p. 107 for observation suggestions for the first five activities. Questions to consider when observing students as they work on Fold and Cut are offered on p. 112.

Fold and Cut

- Do students start by cutting, or do they draw a pattern before cutting?
- How do they react if their results don't match their expectations? Are they able to use their results to modify their next attempt?

Encourage students to talk about both parts of the paper having the same shape. Some may bring up symmetry or balance—a cut on one side of the paper also appears on the other side. Students might also notice that the shapes are in opposite directions.

Suggest to students that they investigate which geometric shapes (squares, rectangles, some triangles, and so on) have more than one line of symmetry. They can do this by folding each shape.

Activity

Class Discussion: Is It Symmetrical?

About 20 minutes before the end of Session 6, students can share the designs they made with color tiles and pattern blocks. As students show a design, they should describe why it is symmetrical. Ask them to explain how they planned and made their designs and why they think the designs are symmetrical. Do not require students to use the terms *symmetry* and *symmetrical;* encourage them to use their own language to explain their thinking.

Talk about some ways to know for sure: **How could you prove this shape is symmetrical? What are some other ways you found for showing symmetry.**

Choose one student's recorded design (with mirror symmetry) and show it to the class. As students watch, fold the design along the line of symmetry so the colors are on the outside (rather than on the inside) of the paper.

One way to see if this design is symmetrical is to pinch a colored square.

Demonstrate how to pinch, bringing forefinger and thumb together in the same place on opposite sides of the paper. Students may want to try pinching their fingers together in the air a few times. Hold the paper so only the top side is visible to students and ask someone to pinch one of the colored squares.

Helena, pinch one of the colored squares on this folded paper. What color did you pinch under your top finger? What color do you think is under your thumb on the bottom side of the paper?

Students check their color tile designs by folding and pinching colors. Suggest they check their designs using another method, such as a mirror.

Symmetrical Pictures

What Happens

Students make pictures using symmetrical shapes cut from folded paper. They write an explanation of how they know their shapes are symmetrical and define line symmetry. Their work focuses on:

- planning how symmetrical shapes can be used to make a picture
- making symmetrical shapes by folding and cutting paper patterns
- writing about symmetry

Start-Up

Today's Number Suggest that students use multiples of 5 and 10 in their number sentences. For example, if the number they are working on is 105, possible expressions include: 50 + 50 + 5 or 75 + 25 + 5. Add a card to the class counting strip and fill in the next number on the blank 200 chart.

Quick Images Use the Rectangular Arrays images. Provide enough time between flashes for students to figure out the number of squares in the array. Have students share how they figured out the number.

Materials

- Letter-size paper, half sheets (3 per student)
- Construction paper (1–3 per student)
- Scissors
- Paste or glue sticks
- Crayons or markers
- Student Sheet 27 (1 per student, homework)
- Chart paper (optional)

Activity

As a final project, small groups of students can work together to make pictures using symmetrical cutouts. Construction paper can be used for the background. Everything in the picture must be symmetrical and made by cutting a folded piece of paper or by combining two or more cutout shapes. (However, the picture as a whole need not be symmetrical.) For example, one student may make a house by folding a piece of paper and cutting out a house shape. Another student may fold and cut out a rectangle and then fold and cut out a triangle to be placed on top of the rectangle as the roof. Encourage students to revise their work if they are dissatisfied with the first attempt. Students should check each other's cutouts to be sure that they are in fact symmetrical.

When students have cut the symmetrical shapes for their pictures, they paste them on construction paper. Students may add details with crayons or markers.

When they have finished, students write two or three sentences explaining how the things in their picture are symmetrical and what it means for something to have line or mirror symmetry. Post students' pictures and explanations.

Assessment

Symmetrical Pictures

The following are some things you may want to think about as you look at students' work.

- How have students made the objects for their pictures? Have they cut out different shapes and put them together to make an object, such as a house, or have they cut out objects as wholes? Have students cut things out by folding paper in half and cutting, or have they tried to cut out a symmetrical object in its entirety?

- Are the objects in their pictures symmetrical? If students have drawn in parts, such as windows, are they also symmetrical? If the drawings are not in the center, are they drawn in positions to balance? If asked, can students identify objects in their pictures that are not symmetrical?

- How have students described their symmetrical pictures in writing? How do they express what symmetry is? Do their explanations match their pictures?

When students are finished, they share their work and discuss how the pictures are symmetrical.

Jeffrey
This is a symmetrecl Birthday party! I did fold and cut on evrey thing. You know if someting is symmetrecl if it is the same on each side.

Paul
Symytry is when something is the same on each side. The butterfly and boat is symetryele.

As the unit ends, you may want to use one of the following options for creating a record of students' work on this unit.

■ Students look back through their folders and think about what they learned in this unit, what they remember most, what was hard or easy for them. You might have students discuss this with partners or have them share ideas with the whole class, while you record their responses on chart paper.

■ Depending on how you organize and collect student work, you may want to have students select some examples of their work to keep in a math portfolio. In addition, you may want to choose some examples from each student's folder to include. Items such as Today's Number, their initial writing about What Is a Rectangle? the Picturing Rectangles assessment, their Half-and-Half Flags, and their Symmetrical Pictures can be useful pieces for assessing student growth over the school year. You may want to keep the original and make copies of these pieces for students to take home.

■ Send a selection of work home for families to see. Students can write a cover letter describing their work in this unit. This work should be returned if you are keeping a year-long portfolio of mathematics work for each student.

Choosing Student Work to Save

Session 7 Follow-Up

Fold and Cut   Students continue to explore the fold-and-cut activity at home. The directions are on Student Sheet 27, Fold and Cut. You may decide to mount and display some of the shapes students create.

🏠 Homework

Today's Number

Today's Number is one of the routines that are built into the grade 2 *Investigations* curriculum. Routines provide students with regular practice in important mathematical ideas such as number combinations, counting and estimating data, and concepts of time. For Today's Number, which is done daily (or most days), students write equations that equal the number of days they have been in school. Each day, the class generates ways to make that number. For example, on the tenth day of school, students look for ways to combine numbers and operations to make 10.

This routine gives students an opportunity to explore some important ideas in number. By generating ways to make the number of the day, they explore:

- number composition and part-whole relationships (for example, 10 can be 4 + 6, 5 + 5, or 20 – 10)
- equivalent arithmetical expressions
- different operations
- ways of deriving new numerical expressions by systematically modifying prior ones (for example, 5 + 5 = 10, so 5 + 6 = 11)

Students' strategies evolve over time, becoming more sophisticated as the year progresses. Early in the year, second graders use familiar numbers and combinations, such as 5 + 5 = 10. As they become accustomed to the routine, they begin to see patterns in the combinations and have favorite kinds of number sentences. Later in the year, they draw on their experiences and increased understanding of number. For example, on the forty-ninth day they might include 100 – 51, or even 1000 – 951 in their list of ways to make 49. The types of number sentences that students contribute over time can provide you with a window into their thinking and their levels of understanding of numbers.

If you are doing the full-year grade 2 curriculum, Today's Number is introduced in the first unit, *Mathematical Thinking at Grade 2*.

Throughout the curriculum, variations are often introduced as whole-class activities and then carried on in the Start-Up section. The Start-Up section at the beginning of each session offers suggestions of variations and extensions of Today's Number.

While it is important to do Today's Number every day, it is not necessary to do it during math time. In fact, many teachers have successfully included Today's Number as part of their regular routines at the beginning or end of each day. Other teachers incorporate Today's Number into the odd 10 or 15 minutes that exist before lunch or before a transition time.

If you are teaching an *Investigations* unit for the first time, rather than using the number of days you have been in school as Today's Number, you might choose to use the calendar date. (If today is the sixteenth of the month, 16 is Today's Number.) Or you might choose to begin a counting line that does not correspond to the school day number. Each day, add a number to the strip and use this as Today's Number. Begin with the basic activity and then add variations once students become familiar with this routine.

The basic activity is described below, followed by suggested variations.

Materials

- Chart paper
- Student Sheet 1, Weekly Log
- Interlocking cubes

If you are doing the basic activity, you will also need the following materials:

- Index cards (cut in half and numbered with the days of school so far, for example, 1 through 5 for the first week of school)
- Strips of adding-machine tape
- Blank 200 charts (tape two blank 100 charts together to form a 10-by-20 grid)

Continued on next page

Basic Activity

Initially, you will want to use Today's Number in a whole group, starting the first week of school. After a short time, students will be familiar with the routine and be ready to use it independently.

Establishing the Routine

Step 1. Post the chart paper. Call students' attention to the small box on their Weekly Logs in which they have been recording the number of days they have been in school.

Step 2. Record Today's Number. Write the number of the day at the top of the chart paper. Ask students to suggest ways of making that total.

Step 3. List the number sentences students suggest. Record their suggestions on chart paper. As you do so, invite the group to confirm each suggestion or discuss any incorrect responses, and to explain their thinking. You might have interlocking cubes available for students to double-check number sentences.

Step 4. Introduce the class counting strip. Show students the number cards you made and explain that the class is going to create a counting strip. Each day, the number of the day will be added to the row of cards. Post the cards in order in a visible area.

Step 5. Introduce the 200 chart. Display the blank chart and explain that another way the class will keep track of the days in school is by filling in the chart. Record the appropriate numbers in the chart. Tell the class that each day the number of the day will be added to the chart. To help bring attention to landmark numbers on the chart, ask questions such as "How many more days until the tenth day of school? the twentieth day?"

Variations

When students are familiar with the structure of Today's Number, you can connect it to the number work they are doing in particular units.

Make Today's Number Ask students to use some of the following to represent the number:

- only addition
- only subtraction
- both addition and subtraction
- three numbers
- combinations of 10 ($23 = 4 + 6 + 4 + 6 + 3$ or $23 = 1 + 9 + 2 + 8 + 3$)
- a double ($36 = 18 + 18$ or $36 = 4 + 4 + 5 + 5 + 9 + 9$)
- multiples of 5 and 10 ($52 = 10 + 10 + 10 + 10 + 10 + 2$ or $52 = 5 + 15 + 20 + 10 + 2$)

Introduce the idea of working backward. Put the number sentences for Today's Number on the board and ask students to determine what number you are expressing: $10 + 3 + 5 + 7 + 5 + 4 = ?$ Notice how students add this string of numbers. Do they use combinations of 10 or doubles to help them?

In addition to defining how Today's Number is expressed, you can vary how and when the activity is done:

Start the Day with Today's Number Post the day's chart paper ahead of time. When students begin arriving, they can generate number sentences and check them with partners, then record their ways to make the number of the day before school begins. Students can review the list of ways to make the number at that time or at the beginning of math class. At whole-group meeting or morning meeting, add the day's number to the 200 chart and the counting strip.

Continued on next page

Choice Time Post chart paper with the Number of the Day written on it so that it is accessible to students. As one of their choices, students generate number sentences and check them with partners, then record them on the chart paper.

Work with a Partner Each student works with a partner for 5 to 10 minutes and lists some ways to make the day's number. Partners check each other's work. Pairs bring their lists to the class meeting or sharing time. Students have their lists of number sentences in their math folders. These can be used as a record of students' growth in working with number over the school year.

Homework Assign Today's Number as homework. Students share number sentences sometime during class the following day.

Catch-Up It can be easy to get a few days behind in this routine, so here are two ways to catch up. Post two or three Number-of-the-Day pages for students to visit during Choice Time or free time. Or assign a Number of the Day to individual students. Each can generate number sentences for his or her number as well as collect number sentences from classmates.

Class History Post "special messages" below the day's number card to create a timeline about your class. Special messages can include birthdays, teeth lost, field trips, memorable events, as well as math riddles.

Today's Number Book Collect the Today's Number charts in a *Number-of-the-Day Book*. Arrange the pages in order, creating chapters based on 10's. Chapter 1, for example, is ways to make the numbers 1 through 10, and combinations for numbers 11–20 become Chapter 2.

How Many Pockets?

How Many Pockets? is one of the classroom routines presented in the grade 2 *Investigations* curriculum. Routines provide students with regular practice in important mathematical ideas such as number combinations, counting and estimating data, and concepts of time. In How Many Pockets? students collect, represent, and interpret numerical data about the number of pockets everyone in the class is wearing on a particular day. This routine often becomes known as Pocket Day. In addition to providing opportunities for comparison of data, Pocket Days provide a meaningful context in which students work purposefully with counting and grouping. Pocket Day experiences contribute to the development of students' number sense—the ability to use numbers flexibly and to see relationships among numbers.

If you are doing the full-year grade 2 *Investigations* curriculum, collect pocket data at regular intervals throughout the year. Many teachers collect pocket data every tenth day of school.

The basic activity is described below, followed by suggested variations. Variations are introduced within the context of the *Investigations* units. If you are not doing the full grade 2 curriculum, begin with the basic activity and then add variations when students become familiar with this routine.

Materials

- Interlocking cubes
- Large jar
- Large rubber band or tape
- Hundred Number Wall Chart and number cards (1–100)
- Pocket Data Chart (teacher made)
- Class list of names
- Chart paper

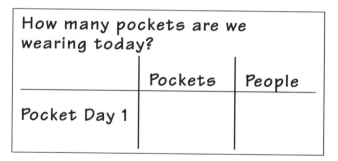

1	2	3	4	5	6	7	8	9	10
11	12	13	14	15	16	17	18	19	20
21	22	23	24	25	26	27	28	29	30
31	32	33	34	35	36	37	38	39	40
41	42	43	44	45	46	47	48	49	50
51	52	53	54	55	56	57	58	59	60
61	62	63	64	65	66	67	68	69	70
71	72	73	74	75	76	77	78	79	80
81	82	83	84	85	86	87	88	89	90
91	92	93	94	95	96	97	98	99	100

Hundred Number Wall Chart

How many pockets are we wearing today?	Pockets	People
Pocket Day 1		

Pocket Data Chart

Basic Activity

Step 1. Students estimate how many pockets the class is wearing today. Students share their estimates and their reasoning. Record the estimates on chart paper. As the Pocket Days continue through the year, students' estimates may be based on the data recorded on past Pocket Days.

Continued on next page

Step 2. Students count their pockets. Each student takes one interlocking cube for each pocket he or she is wearing.

Step 3. Students put the cubes representing their pockets in a large jar. Vary the way you do this. For example, rather than passing the jar around the group, call on students with specific numbers of pockets to put their cubes in the jar (for example, students with 3 pockets). Use numeric criteria to determine who puts cubes in the jar (for example, students with more than 5 but fewer than 8 pockets). Mark the level of cubes on the jar with a rubber band or tape.

Step 4. With students, agree on a way to count the cubes. Count the cubes to find the total number of pockets. Ask students for ideas about how to double-check the count. By recounting in another way, students see that a group of objects can be counted in more than one way, for example, by 1's, 2's, 5's, and 10's. With many experiences, they begin to realize that some ways of counting are more efficient than others and that a group of items can be counted in ways other than by 1, without changing the total.

Primary students are usually most secure counting by 1's, and that is often their method of choice. Experiences with counting and grouping in other ways help them begin to see that number is conserved or remains the same regardless of its arrangement—20 cubes is 20 whether counted by 1's, 2's, or 5's. Students also become more flexible in their ability to use grouping, especially important in our number system, in which grouping by 10 is key.

Step 5. Record the total for the day on a Pocket Data Chart. Maintaining a chart of the pocket data as they are accumulated provides natural opportunities for students to see that data can change over time and to compare quantities.

How many pockets are we wearing today?		
	Pockets	People
Pocket Day 1	41	29

Variations

Comparing Data Students revisit the data from the previous Pocket Day and the corresponding cube level marked on the now-empty jar.

On the last Pocket Day, we counted [*give number*] pockets. Do you think we will be wearing more, fewer, or about the same number of pockets today? Why?

After students explain their reasoning, continue with the basic activity. When the cubes have been collected, invite students to compare the present level of cubes with the previous level indicated by the tape or rubber band and to revise their estimates based on this visual information.

Discuss the revised estimates and then complete the activity. After you add the day's total to the Pocket Data Chart, ask students to compare and interpret the data. To facilitate discussion, build a train of interlocking cubes for today's and the previous Pocket Day's number. As students compare the trains, elicit what the cube trains represent and why they have different numbers of cubes.

Use the Hundred Number Wall Chart Do the basic activity, but this time students choose only one way to count the cubes. Then introduce the Hundred Number Wall Chart as a tool that can be used for counting cubes. This is easiest when done with students sitting on the floor in a circle.

Continued on next page

To check our pocket count, we'll put our cubes in the pockets on the chart. A pocket can have just one cube, so we'll put one cube in number 1's pocket, the next cube in number 2's pocket, and keep going in the same way. How many cubes can we put in the first row?

Students will probably see that 10 cubes will fill the first row of the chart.

One group of 10 cubes fits in this row. What if we complete the second row? How many rows of the chart do you think we will fill with the cubes we counted today?

Encourage students to share their thinking. Then have them count with you and help to place the cubes one by one in the pockets on the chart. When finished, examine the chart together, pointing out the total number of cubes in it and the number of complete rows. For each row, snap together the cubes to make a train of 10. As you do so, use the rows to encourage students to consider combining groups of 10. Record the day's total on your Pocket Data Chart.

Note: If cubes do not fit in the pockets of the chart, place the chart on the floor and place the cubes on top of the numbers.

Find the Most Common Number of Pockets
Each student connects the cubes representing his or her pockets into a train. Before finding the total number of pockets, sort the cube trains with students to find the most common number of pockets. Pose and investigate additional questions, such as:

- **How many people are wearing the greatest number of pockets?**
- **Is there a number of pockets no one is wearing?**
- **Who has the fewest pockets?**

The cubes are then counted to determine the total number of pockets.

Take a Closer Look at Pocket Data Each student builds a cube train representing his or her pockets. Beginning with those who have zero

pockets, call on students to bring their cube trains to the front of the room. Record the information in a chart, such as the one shown here. You might make a permanent chart with blanks for placing number cards.

_0_ people have 0 pockets.	_0 pockets_
_4_ people have 1 pocket.	_4_ pockets
_2_ people have 2 pockets.	_4_ pockets
_2_ people have 3 pockets.	_6 pockets_

Pose questions about the data, such as "Two people have 2 pockets. How many pockets is that?" Then record the number of pockets.

To work with combining groups, you might keep a running total of pockets as data are recorded in the chart until you have found the cumulative total.

We counted [12] pockets, and then we counted [6] pockets. How many pockets have we counted so far? Be ready to tell us how you thought about it.

As students give their solutions, encourage them to share their mental strategies. Alternatively, after all data have been collected, students can work on the problem of finding the total number of pockets.

Graph Pocket Data Complete the activity using the variation Find the Most Common Number of Pockets. Leave students' cube trains intact. Each student then creates a representation of the day's pocket data. Provide a variety of materials including stick-on notes, stickers or paper squares, markers and crayons, drawing paper, and graph paper for students to use.

Continued on next page

These cube trains represent how many pockets people are wearing today. Suppose you want to show our pocket data to your family, friends, or students in another classroom. How could you show our pocket data on paper so that someone else could see what we found out about our pockets today?

By creating their own representations, students become more familiar with the data and may begin to develop theories as they consider how to communicate what they know about the data to an audience. Students' representations may not be precise; what's important is that the representations enable them to describe and interpret their data.

Compare Pocket Data with Another Class
Arrange ahead of time to compare pocket data with a fourth- or fifth-grade class. Present the following question to students:

Do you think fifth-grade students wear more, fewer, or about the same number of pockets as second-grade students? Why?

Discuss students' thinking. Then investigate this question by comparing your data with data from another classroom. One way to do this is to invite the other class to participate in your Pocket Day. Do the activity first with the second graders, recording on the Pocket Data Chart how many people have each number of pockets and finding the total number of pockets. Repeat with the other students, recording their data on chart paper. Then compare the two sets of data.

How does the number of pockets in the fifth grade compare to the number of pockets in second grade?

Discuss students' ideas.

Calculate the Total Number of Pockets Divide students into groups of four or five. Each group determines the total number of pockets being worn by the group. Data from each small group are shared and recorded on the board. Using this information, students work in pairs to determine the total number of pockets worn by the class. As a group, they share strategies used for determining the total number of pockets.

In another variation, students share individual pocket data with the group. Each student records this information using a class list of names to keep track. They then determine the total number of pockets worn by the students in the class. Observe how students calculate the total number of pockets. What materials do they use? Do they group familiar numbers together, such as combinations of 10, doubles, or multiples of 5?

Time and Time Again

Time and Time Again is one of the classroom routines included in the grade 2 *Investigations* curriculum. This routine helps students develop an understanding of time-related ideas such as sequencing of events, the passage of time, duration of time periods, and identifying important times in their day.

Because many of the ideas and suggestions presented in this routine will be incorporated throughout the school day and into other parts of the curriculum, we encourage teachers to use this routine in whatever way meets the needs of their students and their classroom. We believe that learning about time and understanding ideas about time happen best when activities are presented *over* time and have relevance to students' experiences and lives.

Daily Schedule Post a daily schedule. Identify important times (start of school, math, music, recess, reading) using both analog (clockface) and digital (10:15) representations. Discuss the daily schedule each day and encourage students to compare the actual starting time of, say, math class with what is posted on the schedule.

Talk Time Identify times as you talk with students. For example, "In 15 minutes we will be cleaning up and going to recess." Include specific times and refer to a clock in your classroom: "It is now 10:15. In 15 minutes we will be cleaning up and going out to recess. That will be at 10:30."

Timing One Hour Set a timer to go off at 1-hour intervals. Choose a starting time and write both the analog time (use a clockface) and the digital time. When the timer rings, record the time using analog and digital times. At the end of the day, students make observations about the data collected. Initially you'll want to use whole and half hours as your starting points. Gradually you can use times that are 10 or 20 minutes after the hour and also appoint students to be in charge of the timer and of recording the times.

Timing Other Intervals Set a timer to go off at 15-minute intervals over a period of 2 hours. Begin at the hour and after the data have been collected, discuss with students what happened each time 15 minutes was added to the time (11:00, 11:15, 11:30, 11:45). You can also try this with 10-minute intervals.

Home Schedule Students make a schedule of important times at home. They can do this both for school days and for nonschool days. They should include both analog and digital times on their schedules. Later in the year they can use this schedule to see if they were really on time for things like dinner, piano lessons, or bedtime. They record the actual time that events happened and calculate how early or late they were. Students can illustrate their schedules.

Comparing Schedules Partners compare important times in their day, such as what time they eat dinner, go to bed, get up, leave for school. They can compare whether events are earlier or later, and some students might want to calculate how much earlier or later these events occur.

Life Timelines Students create a timeline of their life. They interview family members and collect information about important developmental milestones such as learning to walk, first word, first day of school, first lost tooth, and important family events. Students then record these events on a timeline that is a representation of the first 7 or 8 years of their lives.

Clock Data Students collect data about the types of clocks they have in their home—digital or analog. They make a representation of these data and as a class compare their results.

■ **Are there more digital or analog clocks in your house?**

■ **Is this true of our class set of data?**

■ **How could we compare our individual data to a class set of data?**

Continued on next page

Time Collection Students bring in things from home that have to do with time. Include digital and analog clocks as well as timers of various sorts. These items could be sorted and grouped in different ways. Some students may be interested in investigating different types of timepieces such as sundials, sand timers, and pendulums.

How Long Is a Minute? As you time 1 minute, students close their eyes and then raise their hands when they think a minute has gone by. Ask, "Is a minute longer or shorter than you imagined?" Repeat this activity or have students do this with partners. You can also do this activity with a half-minute.

What Can You Do in a Minute? When students are familiar with timing 1 minute, they work in pairs and collect data about things they can do in 1 minute. Brainstorm a list of events that students might try. Some ideas that second graders have suggested include writing their names; doing jumping jacks or sit-ups; hopping on one foot; saying the ABC's; snapping together interlocking cubes; writing certain numbers or letters (this is great practice for working on reversals); and drawing geometric shapes such as triangles, squares, or stars. Each student chooses four or five activities to do in 1 minute. Before they collect the data, they predict how many they can do in 1 minute. Then with partners they gather the data and compare.

How Long Does It Take? Using a stopwatch or a clock with a second hand, time how long it takes students to complete certain tasks such as lining up, giving out supplies, or cleaning up after math time. Emphasize doing these things in a responsible way. Students can take turns being "timekeepers."

Stopwatches Most second graders are fascinated by stopwatches. You will find that students come up with many ideas about what to time. If possible, acquire a stopwatch for your classroom. (Inexpensive ones are available through educational supply catalogs.) Having stopwatches available in the classroom allows students to teach each other about time and how to keep track of time.

Quick Images

Quick Images is one of the classroom routines in the grade 2 *Investigations* curriculum. Quick Images is introduced in the Geometry and Fractions unit and can be continued through the rest of the year.

This routine can be used along with Today's Number and How Many Pockets? Like those routines, Quick Images should be done on a regular basis. However, the intervals at which the routines are done vary. Today's Number is a daily activity. How Many Pockets? is generally done every 10 days. Consider using Quick Images about two times a week.

In this routine, students are shown an arrangement of dots for a few seconds. Then with the arrangement hidden, students are asked to remember how many dots they saw and tell what strategies they used to help them remember. Students may be asked to draw the arrangement.

This activity encourages students to:

■ develop mental images of spatial representations of number

■ describe their mental images of number

■ attend to the structure of arrays representing number

■ build a visual image of 10

■ see numbers in terms of 10

Materials

■ Overhead projector

■ Transparencies of Quick Images

■ Markers

Note: If an overhead projector is unavailable, draw the images on heavy paper to make Quick Image Cards.

Basic Activity

Prepare the transparencies of Quick Images: Dot Patterns, 10 Frames, Dot Arrays, and Rectangular Arrays. Cut apart the images on the transparencies and store each set in an envelope or resealable plastic bag. Make two copies each of Dot Patterns and 10 Frames so that you can use these sets flexibly—as single numbers or in combinations.

Begin with the Dot Patterns. Briefly describe the activity to the students.

I'm going to quickly show you a picture of some dots. Then I'll cover the picture, and I'll ask you to tell me how many dots there are.

Step 1. Flash an image for 3 seconds. Begin with arrangements of dots for small numbers. For example, start with images of five or six dots arranged as they are on dice or dominoes. Show the image for 3 seconds. (If you show it for too long, students will work from the picture rather than with an mental image. If you show it too briefly, students will not have time to form a mental image.) Encourage them to look at the dot arrangement carefully while it is visible.

Step 2. Students tell how many dots they saw. With the picture covered (or Quick Image card hidden from view), ask students to tell how many dots there are and to describe what they saw.

Step 3. Flash the image again for 3 seconds, then hide it again. This gives students an opportunity to adapt their visual images.

Step 4. Ask students how many dots they think there are. Once students answer, ask them to describe what they saw.

Step 5. Display the pattern one more time. Leave it visible and ask students how many dots there are. Encourage them to explain how they reached their answer.

Continued on next page

Variations

Drawing Dots   Once students are familiar with Quick Images, have them draw the dots after the first time they see the image. When students have finished drawing, quickly show the image again and ask them to make any necessary revisions to their drawing. Then display the image and ask students, "Is your drawing the same as this picture of dots? If not, how is it different?"

Dots Arranged in Arrays   When students have had experiences with visualizing larger groups that are based on dice patterns (for example, 4 groups of five dots), introduce dots arranged in arrays. For example, you might show 4 columns of 3 dots, or 4 columns of 4 dots.

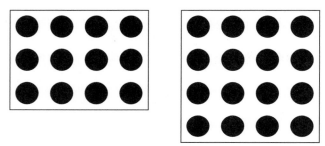

Follow the basic procedure. After you show an array for 3 seconds, students should *draw* the dots and tell you what they saw. Then display the dot array again and have students compare their drawings to it.

Dots in a 10 Frame   Use the basic procedure, but show dots arranged in a 10 frame. After covering the array, ask students how many dots there are and what they saw. When you display the array, ask, "How many dots are there? How did you figure that out?" Using 10 frames helps students to establish 10 as a unit, and to use it as a benchmark—seeing numbers in terms of ten.

Note: If you display the 10 frames vertically, place the card so the left row is filled first, as in the illustration. If you display the 10 frames horizontally, the top row should be filled first.

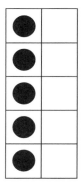

Dots in Two 10 Frames   When students are comfortable with Quick Images using one 10 frame, present arrays in two 10 frames. Once students have a visual image of 10, they are better able to recompose a number in terms of 10. Show the picture below to find if students can see this example as 10—the two 10 frames make a single, filled frame.

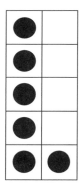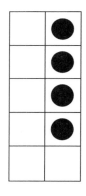

The following activities will help ensure that this unit is comprehensible to students who are acquiring English as a second language. The suggested approach is based on *The Natural Approach: Language Acquisition in the Classroom* by Stephen D. Krashen and Tracy D. Terrell (Alemany Press, 1983.) The intent is for second-language learners to acquire new vocabulary in an active, meaningful context.

Note that *acquiring* a word is different from *learning* a word. Depending on their level of proficiency, students may be able to comprehend a word upon hearing it during an investigation without being able to say it. Other students may be able to use the word orally but not read or write it. The goal is to help students naturally acquire targeted vocabulary at their present level of proficiency.

We suggest using these activities just before the related investigations. The activities can be led by English-proficient students.

Investigations 1-3

shape, fewest

1. Cut the following shapes from construction paper: 5 triangles, 3 rectangles, and 2 squares. Arrange shapes in three groups: triangles, rectangles, and squares.

2. Show each of the groups to students and identify the shapes. Ask students to count and tell how many are in each group.

3. Ask students to point to the group with the fewest number of shapes.

4. Rearrange the shapes into various groupings. Ask students to identify the group with the fewest number of shapes each time.

predict, cover

1. Show students one of the construction paper triangles and a pile of paper clips. Ask students to predict how many paper clips are needed to cover the triangle.

2. Ask a volunteer to place paper clips on the triangle one at a time as the rest of the group counts aloud. Compare some of the predictions and the actual count.

3. Repeat the process of predicting and covering the shape with paper clips for the square and rectangle.

design, match

1. Draw the same design such as polka dots or stripes inside two of the paper triangles. Point out that these triangles now match each other but not the rest of the paper shapes.

2. Ask students to draw matching shapes on paper. Make sure they include a design in each shape.

stories, floor, buildings

1. Show students pictures of buildings in books or magazines.

2. Draw a simple picture of three buildings on the chalkboard. Make the first building one story, the second building two stories, and the third building six stories. Label each floor of the building.

3. Point to the first building and tell students that it has one story. There are not any rooms above the first floor.

4. Point to the second building and count the number of stories aloud. Tell students that this building has two floors. Emphasize the second story.

5. Point to the third building and count the number of stories aloud. Ask students how many floors the building has.

6. Ask students to draw on paper a building that has more than one story. Then, in turn, each student can ask the rest of the group to guess how many stories he or she has drawn. When the correct guess is made, the student reveals his or her drawing.

Teacher Tutorial
Contents

Overview

The units in *Investigations in Number, Data, and Space*® ask teachers to think in new ways about mathematics and how students best learn math. Units such as *Mathematical Thinking* add another challenge for teachers—to think about how computers might support and enhance mathematical learning. Before you can think about how computers might support learning in your classroom, you need to know what the computer component is, how it works, and how it is designed to be used in the unit. This Tutorial is included to help you learn these things.

The Tutorial is written for you as an adult learner, as a mathematical explorer, as an educational researcher, as a curriculum designer, and finally—putting all these together—as a classroom teacher. Although it includes parallel (and in some cases the same) investigations as the unit, it is not intended as a walk-through of the student activities in the unit. Rather, it is meant to provide experience using the computer program *Shapes* and to familiarize you with some of the mathematical thinking in the unit.

The Tutorial is organized in sessions parallel to the unit. Included in each session are detailed step-by-step instructions for how to use the computer and the *Shapes* program, along with suggestions for exploring more deeply. The later parts of the Tutorial include more detail about each component of *Shapes* and can be used for reference. There is also detailed help available in the *Shapes* program itself.

Shapes is a computer manipulative, a software version of pattern blocks and tangrams, that extends what students can do with these shapes. Students create as many copies of each shape as they want and use computer tools to move, combine, and duplicate these shapes to make pictures and designs and to solve problems.

Teachers new to using computers and *Shapes* can follow the detailed step-by-step instructions. Those with more experience might not need to read each step. As is true with learning any new approach or tool, you will test out hypotheses, make mistakes, be temporarily stumped, go down wrong paths, and so on. This is part of learning but may be doubly frustrating because you are dealing with computers. It might be helpful to work through the Tutorial and the unit in parallel with another teacher. If you get particularly frustrated, ask for help from the school computer coordinator or another teacher more familiar with using computers. It is not necessary to complete the Tutorial before beginning to teach the unit. You can work through in parts, as you prepare for parallel investigations in the unit.

Although the Tutorial will help prepare you for teaching the unit, you will learn most about *Shapes* and how it supports the unit as you work side by side with your students.

About *Shapes*

Shapes is a computer manipulative, a software version of pattern blocks and tangrams, that extends what students can do with these shapes. Students create as many copies of each shape as they want and use computer tools to move, combine, and duplicate these shapes to make pictures and designs and to solve problems.

What Should I Read First?

Read the next section, Starting *Shapes*, for specific information on how to load the *Shapes* program and choose an activity.

The section Free Explore takes you step by step through an example of working with *Shapes*. Read this section for a sense of what the program can do.

The section Using *Shapes* provides detailed information about each aspect of *Shapes*. Read this to learn *Shapes* thoroughly or to answer specific questions.

Starting *Shapes*

Note: These directions assume that *Shapes* has been installed on the hard drive of your computer. If not, see How to Install *Shapes* on Your Computer, p. 163.

1. Turn on the computer by doing the following:

   a. If you are using an electrical power surge protector, switch to the ON position.

   b. Switch the computer (and the monitor, if separate) to the ON position.

   c. Wait until the desktop or workspace appears.

2. Open *Shapes* by doing the following:

   a. Double-click on the *Shapes* Folder icon if it is not already open. To double-click, click twice in rapid succession without moving the pointer.

   b. Double-click on the *Shapes* icon in this folder.

c. Wait until the *Shapes* opening screen appears. Click on the bar "Click on this window to continue." when the message appears.

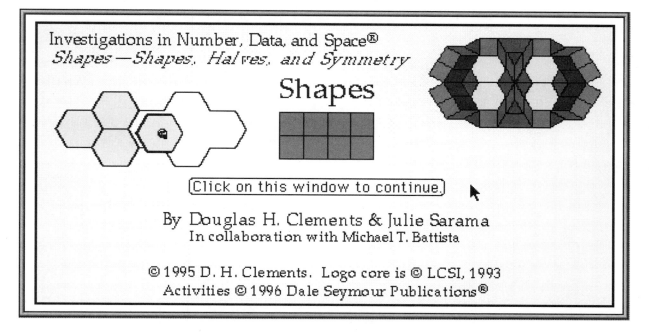

How to Start an Activity

Start an activity by doing the following:

Click on Free Explore (or any activity you want).

Solve Puzzles	Create Your Puzzle	Solve Your Puzzle	Create a Tiling
Mirrors	Open My Work	Free Explore	Click on an activity.

When you choose an activity, the Tool bar, Shape bar, and Work window fill the screen.

The following section provides a step-by-step example of working with *Shapes*.

About Free Explore

The Free Explore activity is available for you to use as an environment to explore *Shapes*. It can also be used to extend and enhance activities.

When you choose Free Explore, you begin with an empty Work window. You can build a picture in that window with the shapes from the Shape bar. The tools in the Tool bar enable you to move, duplicate, and glue the shapes you select from the Work window.

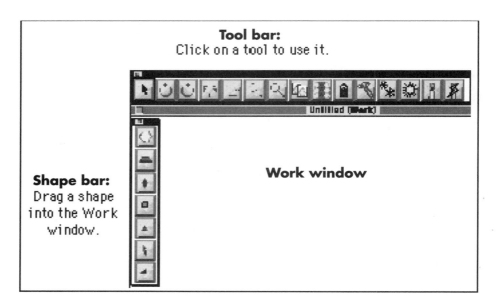

Tool bar:
Click on a tool to use it.

Untitled (Work)

Shape bar:
Drag a shape into the Work window.

Work window

Building a Picture

Let's begin by making a building in the Work window.

 1. Build the front of the building by doing the following:

a. Drag an orange square shape off the Shape bar.

| Move the cursor so it is on the square. It becomes a hand. | Click the mouse button and hold it down . . . | . . . while you move the square where you want it. |

b. Slide the square to the middle of the Work window.

If you need to move the square, just click on it and drag it again.

c. Drag another orange square shape from the Shape bar and place it next to the first one.

Notice that the new square "snaps" right next to the first one.

d. Continue this procedure until the "front" of the building is finished.

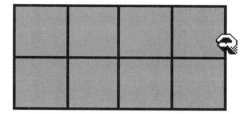

☞ **2.** Build the side of the building by doing the following:

Drag two tan rhombuses (thin diamonds) for the side of the building from the Shape bar and slide into place.

Young students might not do this, but we're going to try for a 3-dimensional effect!

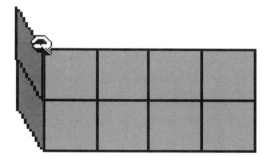

☞ **3.** Start the roof of the building by doing the following:

Drag a blue rhombus (diamond) from the Shape bar.

This shape will have to be turned to make it fit.

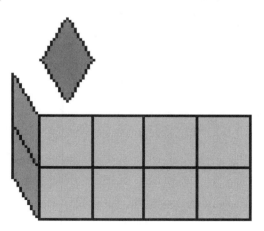

4. Turn the blue rhombus by doing the following:

a. Click on the Left Turn tool in the Tool bar.

The cursor changes from an arrow to a left-turn circle.

b. Click this new cursor on the blue rhombus.
The rhombus turns to the left.

c. Click on the blue rhombus a second time.
The rhombus turns to the left again, and it is ready to be slid into place.

5. Slide the blue rhombus into place by doing the following:

a. Click on the Arrow tool in the Tool bar.
The cursor changes back to an arrow.

b. Slide the rhombus into place.

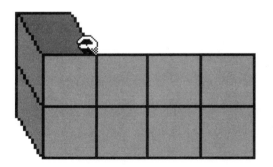

☛ **6.** Duplicate the blue rhombus three times to make three copies of it:

    a. Click on the Duplicate tool in the Tool bar.

        The cursor changes from an arrow to the Duplicate icon .

    b. Click the Duplicate tool *on* the blue rhombus.

        A duplicate is made. Using the Duplicate tool is particularly appropriate in this case because the duplicate is turned the correct way automatically.

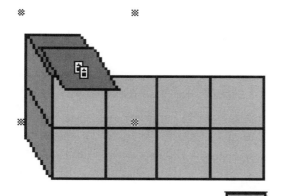

    c. Click on the Arrow tool in the Tool bar.
        The cursor changes back to an arrow.

    d. Slide the duplicate rhombus into place.

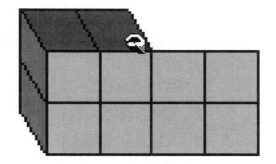

e. Repeat steps a. to d. to make and position two more duplicates.

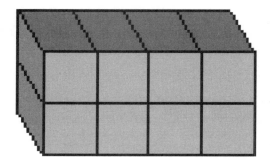

Note: To be more efficient, we could have duplicated three copies right away and then slid each into place one after the other.

☛ **7.** Make a half-circle doorway.

a. Drag a quarter circle from the Shape bar. Slide it into place.

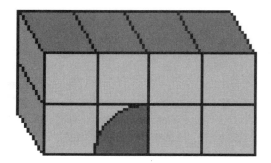

b. Drag another quarter circle (or duplicate the first one) and slide it into place.

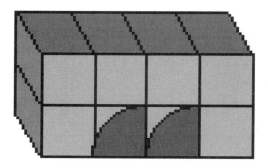

c. Click on the Vertical Flip tool in the Tool bar.

The cursor changes from an arrow to the Vertical Flip icon ⫚.

d. Click the Vertical Flip tool on the second quarter circle.

The shape flips over a vertical line through the center of the shape. Because the quarter circle is symmetric, we could have also turned it several times to the right, but flipping is more efficient.

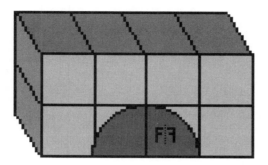

Our building is finished.

☞ **1.** Make a sun.

a. Drag a yellow hexagon from the Shape bar.

Place it in the upper right-hand corner of the Work window.

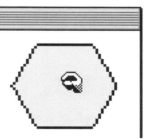

b. Get another yellow hexagon and place it right over the first one.

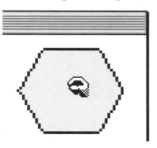

c. Click on the Right Turn tool in the Tool bar.

The cursor changes from an arrow to the Right Turn icon ↻.

More About Building Pictures

d. Click the Right Turn tool on the second hexagon.

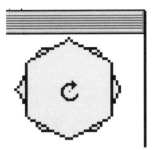

Our sun is finished. Now let's add a walkway. It will be a pattern of several shapes. First, we'll design the unit.

☛ **1.** Make a unit for the walkway.

a. Drag a yellow hexagon, a red trapezoid, a green triangle, and a blue rhombus from the Shape bar.

Place the shapes in the lower left-hand corner of the Work window as shown. Some shapes will have to be turned to make the pattern shown.

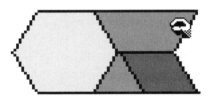

b. Click on the Glue tool in the Tool bar.

The cursor changes from an arrow to the Glue icon 📇.

c. Click in the middle of *each* of the four shapes in the unit.

The cursor changes to a "squirt glue" icon whenever you click on a shape that has not yet been glued. Note that you have to click on each shape; even though they are "snapped" and touching sides, you must indicate each one you want glued together by clicking in the middle of each shape.

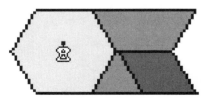

The four shapes are now a glued group. They can be moved, turned, flipped, or duplicated as if they were one shape.

You can check that the shapes are one glued group. Move the cursor to the Glue tool in the Tool bar and hold down the mouse button. The following will appear on your screen, indicating one group.

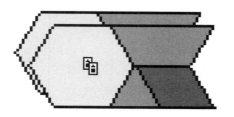

☛ 2. Duplicate the unit for the walkway.

a. Click on the Duplicate tool in the Tool bar.

The cursor changes from an arrow to the Duplicate icon 🔠 .

b. Click the Duplicate tool on the blue rhombus.

A duplicate is made. Using the Duplicate tool is necessary because we're going to make a repeating pattern.

☛ 3. Define the motion for the pattern.

a. Click on the Arrow tool in the Tool bar.
The cursor changes back to an arrow.

b. Slide the duplicate of the unit where you want it to be for the start of the pattern.

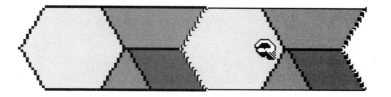

☞ **4.** Continue the pattern.

a. Click on the Pattern button in the Tool bar.

The Pattern tool is a button. Simply clicking on a button will perform the action immediately. The next part of the pattern is put into place.

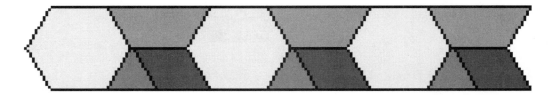

b. Keep clicking on the Pattern button until your walkway extends across the window.

Our walkway is finished. Now let's make trees. Use the Pattern tool again to make the top of the first tree.

☞ **1.** Make a unit for the treetop.

Get a green triangle from the Shape bar.

Place it on the left-hand side of the Work window. We don't need to glue this time because our unit is just one shape.

☞ **2.** Duplicate it.

a. Click on the Duplicate tool in the Tool bar.

The cursor changes from an arrow to the Duplicate icon .

b. Click the Duplicate tool on the triangle.

☞ **3.** Define the motion for the pattern.

a. Click on the Right Turn tool in the Tool bar.

The cursor changes back to the Right Turn icon ↻ .

b. Turn the duplicate of the unit to the right two times.

c. Click on the Arrow tool in the Tool bar.
The cursor changes back to an arrow.

d. Slide the duplicate of the unit where you want it to be for the start of the pattern.

We have now defined the motion for the pattern.

4. Continue the pattern.

a. Click on the Pattern button in the Tool bar four times to

complete the treetop.

We'll finish the tree.

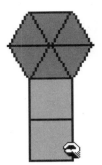

☞ 1. Make a trunk and ground cover.

    a. Drag two squares from the Shape bar. Place the squares below the green triangles as shown.

    b. Drag three tan rhombuses from the Shape bar. Turn them left three times and place them as ground cover. (You could also get one, turn it left three times, then use the Duplicate tool to make two copies.)

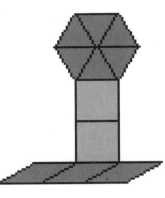

We need ground cover in back of the tree too.

    c. Drag three more tan rhombuses from the Shape bar. Place them in "back" of the others. (You could use the Duplicate tool to do this.)

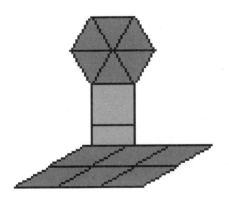

The tree shouldn't be behind the ground cover.

☞ **2.** Bring the tree trunk to the front of the ground cover.

   a. Select the bottom orange square by clicking in the middle of it one time.

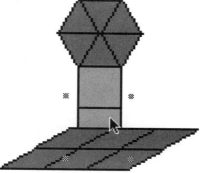

The small gray "selection" squares show that the bottom orange square is "selected." You can choose menu items to apply to selected shapes.

   b. Choose **Bring To Front** from the **Edit** menu.

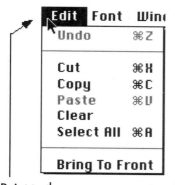

Edit Font Win(	
Undo	⌘Z
Cut	⌘H
Copy	⌘C
Paste	⌘U
Clear	
Select All	⌘A
Bring To Front	

Point to the menu you want and press the mouse button . . .

Edit Font Win(	
Undo	⌘Z
Cut	⌘H
Copy	⌘C
Paste	⌘U
Clear	
Select All	⌘A
Bring To Front	

. . . then move the cursor to **Bring To Front.**

The orange square is brought to the front of the picture.

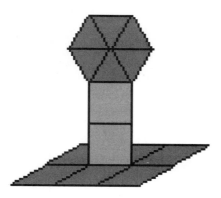

☛ 3. Glue the tree and ground cover together.

In Step 2, you selected a single shape, the orange square, and applied the action (**Bring To Front**) to it. You can also select a *group* of shapes and apply a tool or action to the entire group at one time. This will make gluing all these shapes together easier.

a. Place the arrow at the top left corner of the group of shapes in the tree.

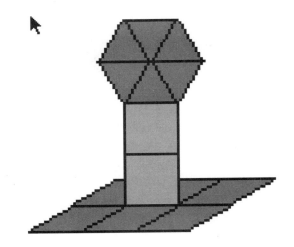

b. Drag diagonally to enclose the shapes in a dotted rectangle . . .

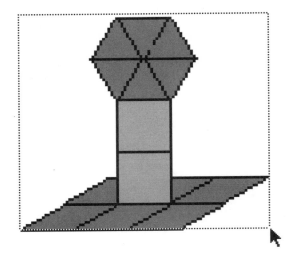

c. . . . and release the mouse button.

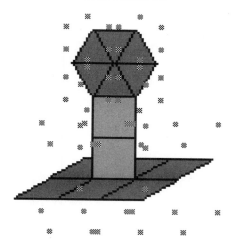

d. Use the Glue tool to glue all the shapes into a group at once by clicking in the middle of any one of them.

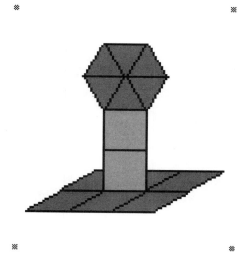

Click on the Arrow tool in the Tool bar. Now the small gray "selection" squares will surround a whole group instead of each individual shape.

You can now do something to all these shapes at once: duplicate them, slide them, turn them, or flip them as one shape. Note that if you click on one of the selected group and slide the whole group, there may be a delay while the *Shapes* program builds an outline of the group.

Next let's try duplicating a group.

 1.   Duplicate the tree.

Use the Duplicate tool to make several copies of the tree and place them where you like.

 2.   Add any finishing touches.

In the picture below, blue rhombuses connecting the building and the walkway were added, some shapes were moved, and the **Bring To Front** command was applied to add a few finishing touches.

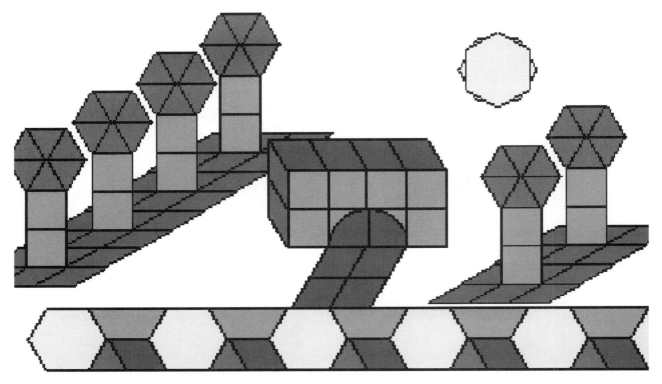

The picture is finished.

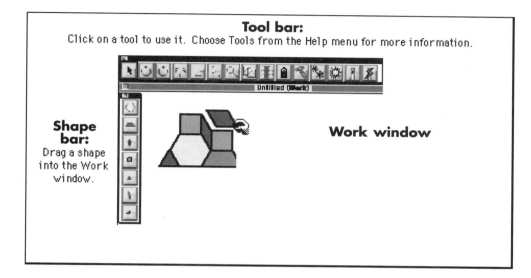

Tool bar:
Click on a tool to use it. Choose Tools from the Help menu for more information.

Shape bar:
Drag a shape into the Work window.

Work window

Tool Bar, Shape Bar, and Work Window

Begin by dragging a shape from the Shape bar (the vertical, "floating" bar on the left) to the Work window (the large blank window). Dragging means clicking on a shape and then holding the mouse button down while you move the mouse.

Move the cursor to a shape with the mouse. It becomes a hand.

Click the mouse button and hold it down . . .

. . . while you move the new shape where you want it.

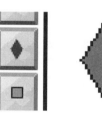

Once the shape is placed in the Work window, you can slide it again by dragging it with the Arrow tool. If you place one shape so that one of its sides is close to a side of another shape, the two shapes will "snap" together.

You can change the position of the shape, or duplicate it, by using the tools in the Tool bar. The tool that is "active," or in use, is surrounded by a black outline (like the arrow tool shown in the diagram on p. 150). Another way to see which tool is active is by the shape of the cursor.

Only the most commonly used tools are available and displayed for each activity. All tools are available for Free Explore.

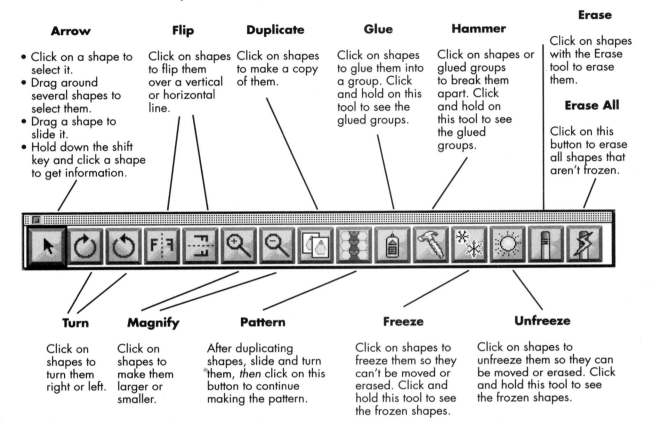

Arrow
- Click on a shape to select it.
- Drag around several shapes to select them.
- Drag a shape to slide it.
- Hold down the shift key and click a shape to get information.

Flip
Click on shapes to flip them over a vertical or horizontal line.

Duplicate
Click on shapes to make a copy of them.

Glue
Click on shapes to glue them into a group. Click and hold on this tool to see the glued groups.

Hammer
Click on shapes or glued groups to break them apart. Click and hold on this tool to see the glued groups.

Erase
Click on shapes with the Erase tool to erase them.

Erase All
Click on this button to erase all shapes that aren't frozen.

Turn
Click on shapes to turn them right or left.

Magnify
Click on shapes to make them larger or smaller.

Pattern
After duplicating shapes, slide and turn them, *then* click on this button to continue making the pattern.

Freeze
Click on shapes to freeze them so they can't be moved or erased. Click and hold this tool to see the frozen shapes.

Unfreeze
Click on shapes to unfreeze them so they can be moved or erased. Click and hold this tool to see the frozen shapes.

To use most tools (except Pattern and Erase All, which are buttons):

☞ 1. Click on a tool in the Tool bar to make it active. The cursor will change to look like the tool.

☞ 2. Click in the middle of a shape to perform the action. If you click one of several shapes that are "selected," the action is performed on each of the selected shapes. See the following section, The Arrow Tool, for more information about selecting shapes.

Pattern and Erase All are **buttons**. Simply clicking on a button will perform the action immediately.

The following sections discuss the tools in more detail.

With the Arrow tool, you can drag shapes to slide them. This is the most important use of the Arrow tool.

1. Click in the middle of a shape and hold the mouse button down . . .

2. . . . while you move the mouse, sliding the shape.

3. Release the button to stop sliding.

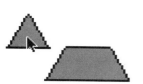

You can also select shapes with the Arrow tool.

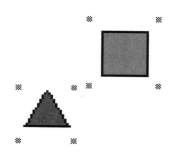

Note: You can do most tasks in *Shapes*, including sliding, without ever selecting shapes. It's usually just a convenience for taking some action on several shapes at once. Before we discuss how to select shapes, let's describe what "selecting" means.

Selected shapes are shown surrounded with small gray squares.

Selected shapes can be copied to the clipboard (a place in computer memory for temporary, invisible storage) or cut—copied to the clipboard *and* removed from the Work window—using commands on the **Edit** menu. Also, if you apply a tool to one shape that is part of a group of selected shapes, the tool will automatically be applied to each shape in the group.

There are two ways to select shapes. First, you can click on any shape with the Arrow tool. That selects the shape. (If it is *already* selected, this will "unselect" the shape.) Second, you can select multiple shapes that are near one another:

1. Place the arrow at one corner of the group of shapes. Press the mouse button.

2. Drag diagonally to enclose the shapes in a dotted rectangle . . .

3. . . . and release the mouse button.

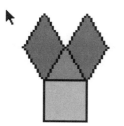 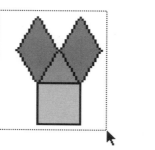 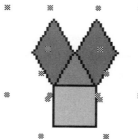

Now you can do something to all these shapes at once; for example, copy or cut them, slide them all together, or flip them all. Note that if you click on one of the selected group and slide the whole group, there may be a delay while the *Shapes* program builds an outline. Hold the mouse button down without moving the mouse until the outline appears.

You can use the Arrow tool to shift-click on a shape to get information about it. To shift-click, hold the shift key down while clicking on a shape. Click again to clear the message.

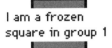

I am a frozen square in group 1

One final feature: If you're using any other tool and you want to use the Arrow tool for a quick selection or slide, just hold down the Command ⌘ key. *Shapes* will know to use the Arrow tool while the Command key is held down. When you let go of the Command key, *Shapes* will return to the previous tool.

Turn, Flip, and Magnifying Tools: Moving and Sizing

You can use these tools to turn or flip shape(s):

One shape: Click on a shape with the tool. For example, if you click on the shape with the first flip tool, the shape flips over a vertical line through the center of the shape.

Several shapes: After the shapes have been selected, click on one of them with the tool. For example, if you click on one with the first Turn tool, each shape turns right around its center.

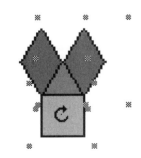

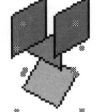

To magnify shapes: If you click on a shape with the first Magnify tool, the shape gets bigger. The second Magnify tool will make it smaller. Shapes that are different sizes will not snap to each other.

You can use the Glue tool to glue several shapes together into a group. This group is a new shape you have created. You can slide, turn, and flip it as a unit—that is, as if it were a single shape. For example, you can glue several shapes and then move them or duplicate them.

To use the Glue tool,

 1. Click on Glue tool in the Tool bar to make it active.

 2. Click on each shape you wish to glue together into a group.

> If there are only two shapes, or if two or more shapes are "snapped" or touching, you still have to click on each of them. Click in the middle of each shape and the computer glues them together. Similarly, if you select a group of shapes, and click on one shape, the group will be glued.

> You can add more shapes to an already glued group. Click on one shape in the glued group, then click on one shape in the new group. The two groups will now be glued.

 3. Click on the Arrow tool or any other tool. All the shapes you glued will become a single, new group.

The small gray "selection" squares will surround a whole group instead of each individual shape.

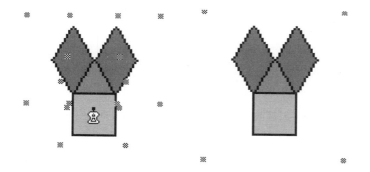

The group will now act as a single unit. For example, if you click on the group with the Right Turn tool, the group turns *as one shape* around the center of the group.

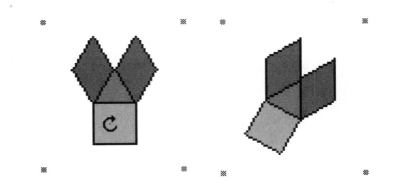

What if you want to make two separate groups? Suppose you made two different kinds of houses and you want each to be glued in a different group. You can't glue all the shapes in each of the houses at once; that would make one two-house group. Instead, you must glue one house, clicking on the Arrow tool to end the gluing process and glue that group. Then you must choose the Glue tool again and glue the second house.

Sometimes it helps to know what shapes are already glued into groups. Hold the mouse button down on either the Glue or Hammer tool on the Tool bar to see which shapes are in which groups.

☞ 1. Hold the button down on the Glue (or Hammer) tool to see the group number on each shape.

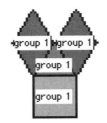

Use the Hammer tool to break apart glued shapes. Click on any shape in the group with the hammer to break apart the group.

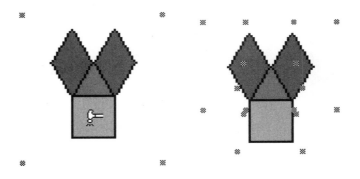

The Pattern tool is a **button**. You just click on it to perform the action (the Erase All button works the same way). To use the Pattern button, you must first make a pattern.

Duplicate Tool and Pattern Button: Making Repeating Patterns

1. Make a basic unit for the pattern. Any shape or combination of shapes can be used as this unit. If you wish to turn the unit later to start the pattern, you must glue the shapes in the basic unit to form a group. Turn patterns must have a single glued group as the basic unit.

2. Duplicate this unit with the Duplicate tool. (You can also choose **Copy** and then **Paste** from the **Edit** menu.)

3. Move the duplicate of the unit where you want it to be for the start of the pattern by using the tools. You can slide the duplicate or turn it. (Remember, if you turn it as in this example, all the shapes in the duplicate must have been glued into one group.)

4. Now, click on the Pattern button as many times as you wish to continue the pattern. If a duplicate goes off the Work window, you will hear a beep and the Pattern button will not make any more copies.

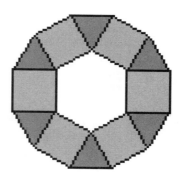

The pattern below was made by sliding the duplicate and then clicking on the Pattern button.

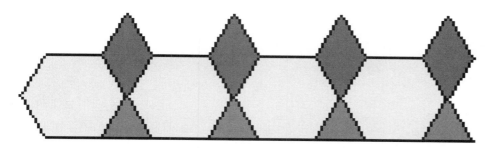

The Pattern tool makes patterns composed of slides and turns. If you want flipped shapes in your pattern, use these steps to make a basic unit (Step 1, p. 155):

a. Glue a group.

b. Duplicate the group.

c. Use the Flip tool to flip the duplicate.

d. Glue the two groups (the original and its flipped [mirror] image) together to make the basic unit for the pattern.

e. Continue with Step 2 (p. 155).

Freeze and Unfreeze Tools: Combining and Breaking Apart

You can use the Freeze tool to "freeze" shapes. Frozen shapes can't be moved or erased. They are frozen in place (in comparison, the Glue tool glues shapes to each other; however, the glued group can still be moved).

Freezing shapes allows you to manipulate other shapes without accidentally moving or erasing those that are frozen. Click on a shape with the Unfreeze tool to unfreeze it. Click and hold either tool on the Tool bar to see the frozen shapes.

Using Menu Commands

To use any menu commands, do the following:

Point to the menu you want and press the mouse button . . .

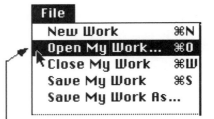

. . . then drag the selection bar to your choice and release the button.

The **File** menu deals with documents and quitting.

New Work starts a new document. The ⌘N indicates that, instead of selecting this from the menu, you could enter ⌘N or **Command N** by holding down the **Command** key (with the ⌘ and ⌘ symbols on it) and then pressing the N key.

Open My Work opens previously saved work.

Close My Work closes present work.

Save My Work saves the work.

Save My Work As saves the work with a new name or to a different disk or folder.

Change Activity lets you choose a different activity.

Page Setup allows you to set up how the printer will print your work.

Print prints your work.

Quit quits *Shapes*.

When you save your work for the first time, a dialogue box opens. Type a name. You may wish to include your name or initials, your work, and the date. For the remainder of that session, you can save your work simply by selecting the **Save** menu item or pressing ⌘S (**Command S**).

To share the computer with others, save your work then choose **Close My Work**. Later, to resume your work, choose **Open My Work** and select the work you saved.

When you **Quit**, you are asked whether you wish to save your work. If you choose to save at that time (you don't have to if you just saved), the same steps are followed.

The **Edit** menu contains choices to use when editing your work.

Cut deletes the selected object and saves it to a space called the clipboard.

Copy copies selected object on the clipboard.

Paste puts the contents of the clipboard into the Work window.

Clear deletes the selected object but does not put it on the clipboard.

Select All selects all shapes. This is not only a fast and handy shortcut, it also helps when some shapes are nearly or totally off the Work window.

Bring To Front puts the selected shapes "in front of" unselected shapes. That is, each new shape you create will be "in front of" the shapes already on the Work window. If you want to change this, select a shape that is in the "back," hidden by other shapes, and choose **Bring To Front**.

About Menus

File	
New Work	⌘N
Open My Work...	⌘O
Close My Work	⌘W
Save My Work	⌘S
Save My Work As...	
Change Activity...	
Page Setup...	
Print...	⌘P
Quit	⌘Q

Edit	
Undo	⌘Z
Cut	⌘H
Copy	⌘C
Paste	⌘U
Clear	
Select All	⌘A
Bring To Front	

The **Font** menu is used to change the appearance of text in the Show Notes window. In order for any command in the **Font** menu to be highlighted, the Show Notes command in the **Windows** menu must be open.

The first names are choices of typeface.

Size and **Style** have additional choices; pull down to select them and then to the right. See the example for **Style** shown at left. The **Size** choice works the same way.

All Large changes all text in all windows to a large-size font. This is useful for demonstrations. This selection toggles (changes back and forth) between **All Large** and **All Small**.

The **Windows** menu shows or hides the windows. If you hide a window such as the Tool bar, the menu item changes to **Show** followed by the name of the window—for example, **Show Tools**. You can also hide a window by clicking in the "close box" in the upper-left corner of the window.

The **Notes** Window is designed to be a word processor. Students might use this to record a strategy they used to solve a problem, or to write notes to themselves as reminders of how they plan to continue the next math session.

The **Shapes** menu contains several commands to change the shapes.

Pattern Blocks and **Tangrams** chooses one of the shape sets for the Shape bar.

All Large, **All Medium**, and **All Small** changes the size of all the shapes. Use the Magnify tools to change the size of some of the shapes. Note that shapes of different sizes do not snap to each other.

The **Number** menu is available for only certain activities that have more than one task. Select a number off the submenu to work on that number task.

The **Options** menu allows you to customize *Shapes*.

Vertical Mirror and **Horizontal Mirror** toggle (turn on and off) the Work windows mirrors, which reflect any action you take.

Snap toggles the "snap" feature, in which shapes, when moved, automatically "snap" or move next to, any other shape they are close to. You may want to turn that feature off if you have lots of shapes on the Work window, or if you are copying many shapes. This feature can slow down moving the shapes. You can turn it off temporarily to speed things up.

The **Help** menu provides assistance.

Windows provides information on the three main windows: the Tool bar, the Shape bar, and the Work window.

Tools provides information on tools (represented on the Tool bar as icons, or pictures).

Directions provides instructions for the present activity.

Hints gives a series of hints on the present activity, one at a time. It is dimmed when there are no available hints.

Number
✓Number 1
Number 2
Number 3
Number 4
Number 5
Number 6
Number 7
Number 8
Number 9
Number 10
Number 11
Number 12
Number 13
Number 14
Number 15

Options	Help
Vertical Mirror	
Horizontal Mirror	
✓Snap	

Help	
Windows...	
Tools...	
Directions...	⌘D
Hints...	⌘H

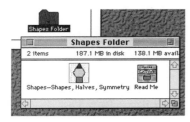

This section contains suggestions for how to correct errors and what to do in some troubling situations.

If you are new to using the computer, you might also ask a computer coordinator or an experienced friend for help.

No *Shapes* Icon to Open

- Check that *Shapes* has been installed on your computer by looking at a listing of the hard disk.
- Open the folder labeled *Shapes* by double-clicking on it.
- Find the icon for the *Shapes* application and double-click on it.

Nothing Happened After Double-Clicking on the *Shapes* Icon

- If you are sure you double-clicked correctly, wait a bit longer. *Shapes* takes a while to open or load and nothing new will appear on the screen for a few seconds.
- On the other hand, you may have double-clicked too slowly, or moved the mouse between your clicks. In that case, try again.

In the Wrong Activity

- Choose **Change Activity** from the **File** menu.

A Window Closed by Mistake

- Choose **Show Window** from the **Windows** menu.

Windows or Tools Dragged to a Different Position by Mistake

- Drag the window back into place by following these steps: Place the pointer arrow in the stripes of the title bar. Press and hold the button as you move the mouse. An outline of the window indicates the new location. Release the button and the window moves to that location.

I Clicked Somewhere and Now *Shapes* Is Gone! What Happened?

You probably clicked in a part of the screen not used by *Shapes* and the computer therefore took you to another application, such as the "desktop."

- Click on a *Shapes* window, if visible.
- Double-click on the *Shapes* program icon.

How Do I Select a Section of Text?

In certain situations, you may wish to copy or delete a section or block of text.

- Point and click at one end of the text. Drag the mouse by holding down the mouse button as you move to the other end of the text. Release the mouse button. Then use the **Edit** menu to **Copy**, **Cut**, and **Paste**.

System Error Message

Some difficulty with the *Shapes* program or your computer caused the computer to stop functioning.

- Turn off the computer and repeat the steps to turn it on and start *Shapes* again. Any work that you saved will still be available to open from your disk.

I Tried to Print and Nothing Happened

- Check that the printer is connected and turned on.
- When printers are not functioning properly, a system error may occur, causing the computer to "freeze." If there is no response from the keyboard or when moving or clicking with the mouse, you may have to turn off the computer and start over.

I Printed the Work Window but Not Everything Printed

- Choose the Color/Grayscale option for printing.
- If your printer has no such option (e.g., an older black and white printer), you need to find a different printer to print graphics in color.

If the *Shapes* program does not understand a command or has a suggestion, a dialogue box may appear with one of the following messages. Read the message, click on **[OK]** or press **<return>** from the keyboard, and correct the situation as needed.

Disk or directory full.

The computer disk is full.

■ Use **Save My Work As** to choose a different disk.

I'm having trouble with the disk or drive.

The disk might be write-protected, there is no disk in the drive, or some similar problem.

■ Use **Save My Work As** to choose a different disk.

Out of space.

There is no free memory left in the computer.

■ Eliminate shapes you don't need.
■ Save and start new work.

The *Shapes* disk that you received with this unit contains the *Shapes* program and a Read Me file. You may run the program directly from this disk, but it is better to put a copy of the program and the Read Me file on your hard disk and store the original disk for safekeeping. Putting a program on your hard disk is called *installing* it.

Note: *Shapes* runs on a Macintosh II computer or above, with 4 MB of internal memory (RAM) and Apple System Software 7.0 or later. (*Shapes* can run on a Macintosh with less internal memory, but the system software must be configured to use a minimum of memory.)

To install the contents of the *Shapes* disk on your hard drive, follow the instructions for your type of computer or these steps:

Slide tab→
up to lock

Back of disk

☞ 1. Lock the *Shapes* program disk by sliding up the black tab on the back, so the hole is open.

   The *Shapes* disk is your master copy. Locking the disk allows copying while protecting its contents.

☞ 2. Insert the *Shapes* disk into the floppy disk drive.

☞ 3. Double-click on the icon of the *Shapes* disk to open it.

☞ 4. Double-click on the Read Me file to open and read it for any recent changes in how to install or use *Shapes*. Click in the close box after reading.

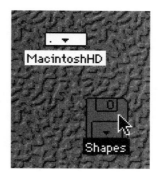

☞ 5. Click on and drag the *Shapes* disk icon (the outline moves) to the hard disk icon until the hard disk icon is highlighted, then release the mouse button.

   The message appears indicating that the contents of the *Shapes* disk are being copied to the hard disk. The copy is in a folder on the hard disk with the name *Shapes*.

☞ 6. Eject the *Shapes* disk by selecting it (clicking on the icon) and choosing **Put Away** from the **File** menu. Store the disk in a safe place.

☞ 7. If the hard disk window is not open on the desktop, open the hard disk by double-clicking on the icon.

   When you open the hard disk, the hard disk window appears, showing you the contents of your hard disk. It might look something like this. Among its contents is the folder labeled *Shapes* holding the contents of the *Shapes* disk.

☛ 8. Double-click the *Shapes* folder to select and open it.

When you open the *Shapes* folder, the window contains the program and the Read Me file.

To select and run *Shapes,* double-click on the program icon.

Optional

For ease at startup, you might create an alias for the *Shapes* program by following these steps:

☛ 1. Select the program icon.

☛ 2. Choose **Make Alias** from the **File** menu.

The alias is connected to the original file that it represents, so that when you open an alias, you are actually opening the original file. This alias can be moved to any location on the desktop.

☛ 3. Move the *Shapes* alias out of the window to the desktop space under the hard disk icon.

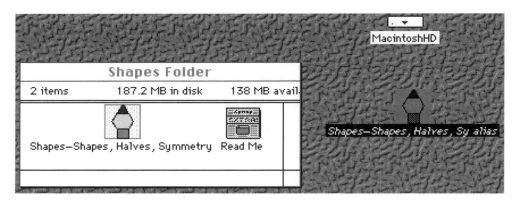

For startup, double-click on the *Shapes* alias instead of opening the *Shapes* folder to start the program inside.

Saving Work on a Different Disk

For classroom management purposes, you might want to save student work on a disk other than the program drive. Make sure that the save-to disk has been initialized (see instructions for your computer system).

☞ 1. Insert the save-to disk into the drive.

☞ 2. Choose **Saue My Work As** from the **File** menu.

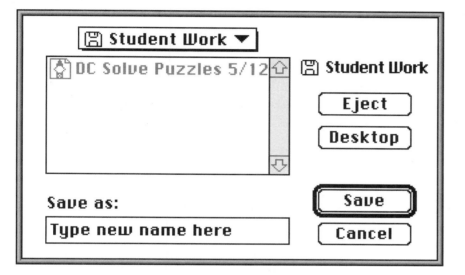

The name of the disk the computer is saving to is displayed in the dialogue box. To choose a different disk, click the **[Desktop]** button and double-click to choose and open a disk from the new menu.

☞ 3. Type a name for your work if you want to have a new or different name from the one it currently has.

☞ 4. Click on **[Save]**.

Deleting Copies of Student Work

As students no longer need previously saved work, you may want to delete their work (called "files") from a disk. This cannot be accomplished from inside the *Shapes* program. However, you can delete files from disks at any time by following directions for how to "Delete a File" for your computer system.

Blackline Masters

_____ , 19 ____

Dear Family,

Our class is beginning a new mathematics unit called *Shapes, Halves, and Symmetry*. For the next few weeks we will be investigating geometry and fractions. We will investigate relationships among shapes and put shapes together to build other shapes. For example, children will use pattern blocks (two-dimensional shapes) to make a hexagon shape by combining two trapezoids or by combining six triangles. Children will look for ways that a large design can be covered with different numbers of blocks.

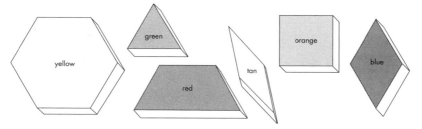

We will also investigate rectangular arrays. Checkerboards and floor tiles are arrays that are familiar to many children. An egg carton is a 2-by-6 array. Activities with arrays provide experiences with fractions and area.

Children will build arrays with square tiles, then draw them. They will compare all the arrays that can be built for a number such as 12. Later in the unit we will use rectangular arrays to show fractions. Each child will design a Fraction Flag that is divided into equal parts.

The last part of this unit is about symmetry. Students will make symmetrical designs with pattern blocks and investigate symmetry using a mirror. These activities, as well as others in this unit, help children develop visual thinking.

You can help your child at home by looking for opportunities to talk about shapes, fractions, and symmetry. For example:

■ Look for different shapes in the environment. Where do you see rectangles and squares? Are there some shapes within other shapes, such as panes in a window?

■ Look for arrays—for example, in floor tiles, calendars, and window panes.

■ Look for patterns in fabric, wallpaper, flags, or other places that are half one color and half another. How can your child tell that the patterns are half and half?

■ Look for designs that are symmetrical.

Have fun exploring these ideas with your child.

Sincerely,

Weekly Log

Monday, _____	
Tuesday, _____	
Wednesday, _____	
Thursday, _____	
Friday, _____	

Shapes at Home

Look around at home and in your neighborhood for shapes. Write about or draw pictures of at least 5 things you found.

Name ______________________________ Date

Student Sheet 3

Predict and Cover (Shapes A and B)

Shape A

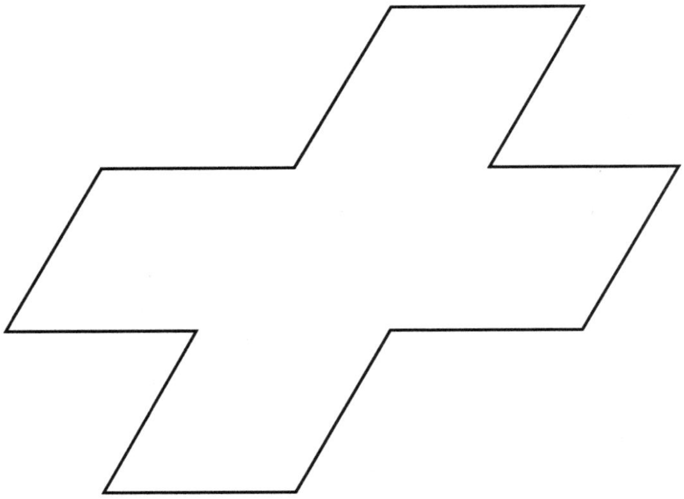

Block: rhombus

Predict: ____________

Count: ____________

Block: triangle

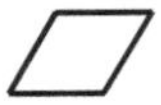

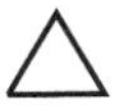

Predict: ____________

Count: ____________

Shape B

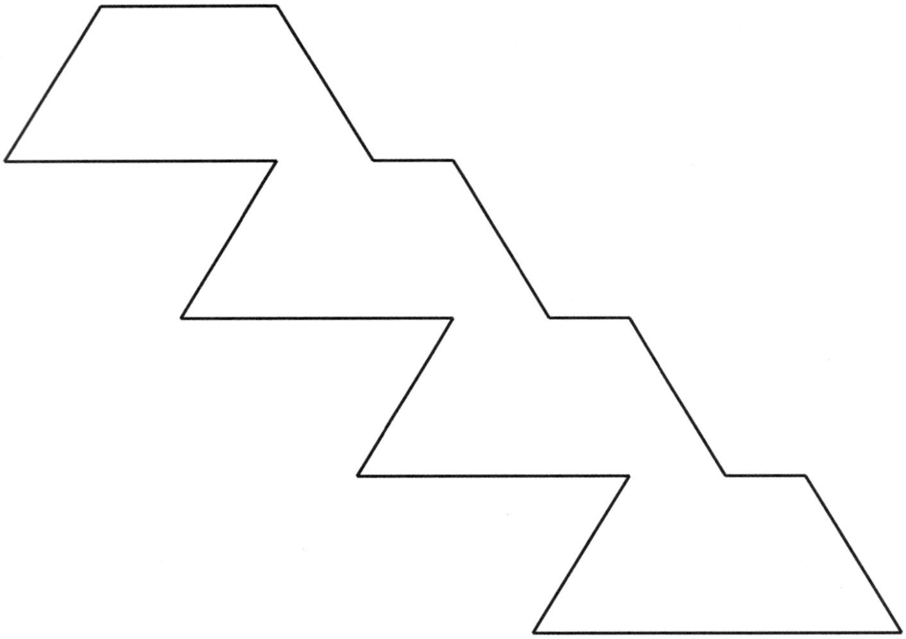

Block: trapezoid

Predict: ____________

Count: ____________

Block: triangle

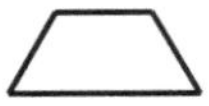

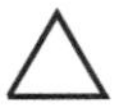

Predict: ____________

Count: ____________

Name _______________________________ Date

Student Sheet 4

Predict and Cover (Shapes C and D)

Shape C

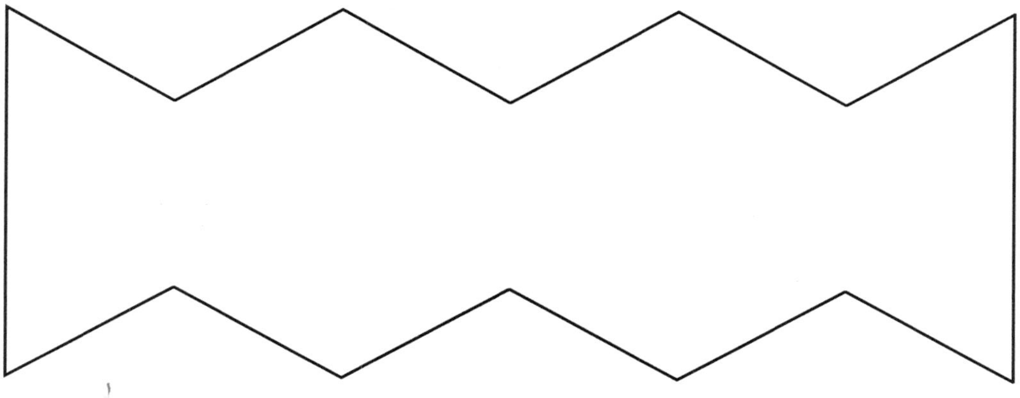

Block: trapezoid
Predict: ____________
Count: ____________

Block: triangle
Predict: ____________
Count: ____________

Shape D

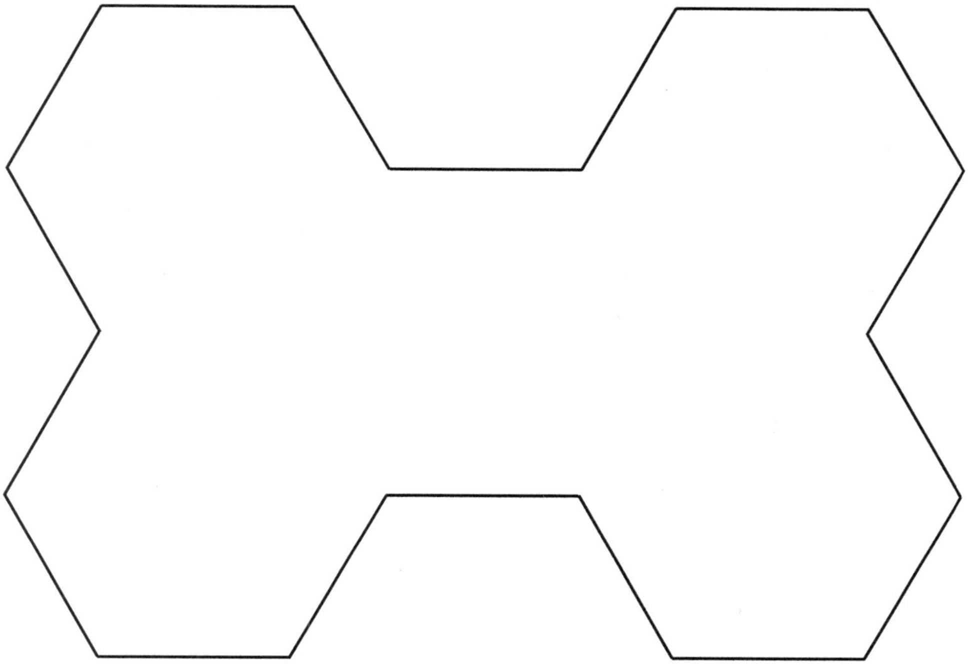

Block: hexagon
Predict: ____________
Count: ____________

Block: trapezoid
Predict: ____________
Count: ____________

Block: triangle
Predict: ____________
Count: ____________

Predict and Cover (Shapes E and F)
Shape E

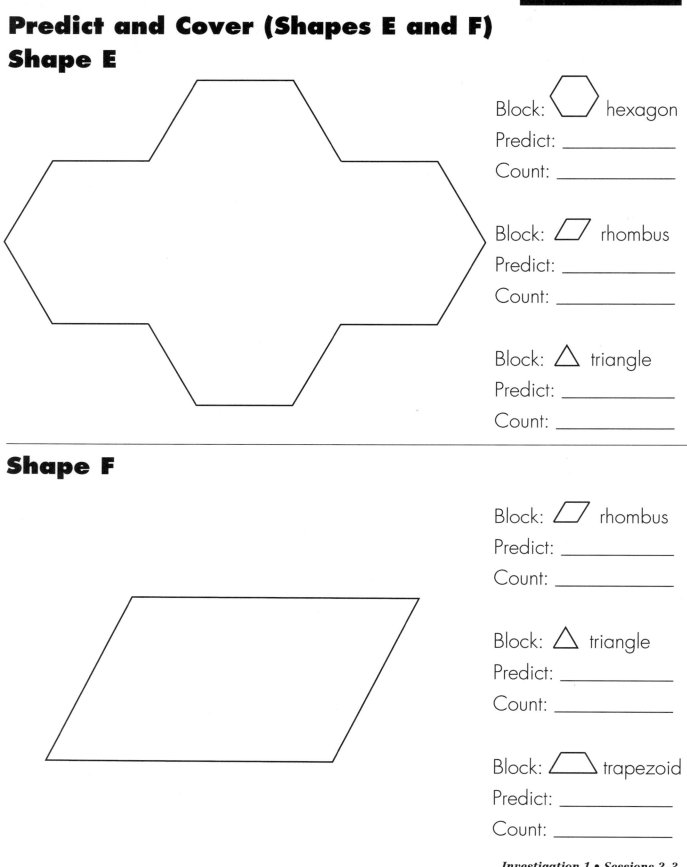

Block: ⬡ hexagon

Predict: _____

Count: _____

Block: ▱ rhombus

Predict: _____

Count: _____

Block: △ triangle

Predict: _____

Count: _____

Shape F

Block: ▱ rhombus

Predict: _____

Count: _____

Block: △ triangle

Predict: _____

Count: _____

Block: ⬭ trapezoid

Predict: _____

Count: _____

Name Date

Student Sheet 6

Predict and Cover (Shapes G and H)

Shape G

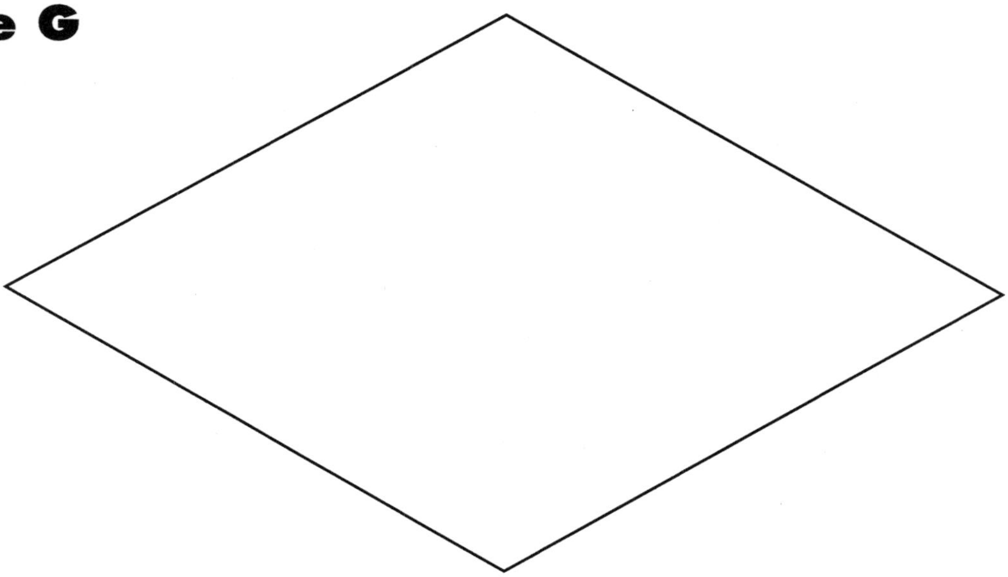

Block: ______________ Block: ______________ Block: ______________
Predict: ______________ Predict: ______________ Predict: ______________
Count: ______________ Count: ______________ Count: ______________

Shape H

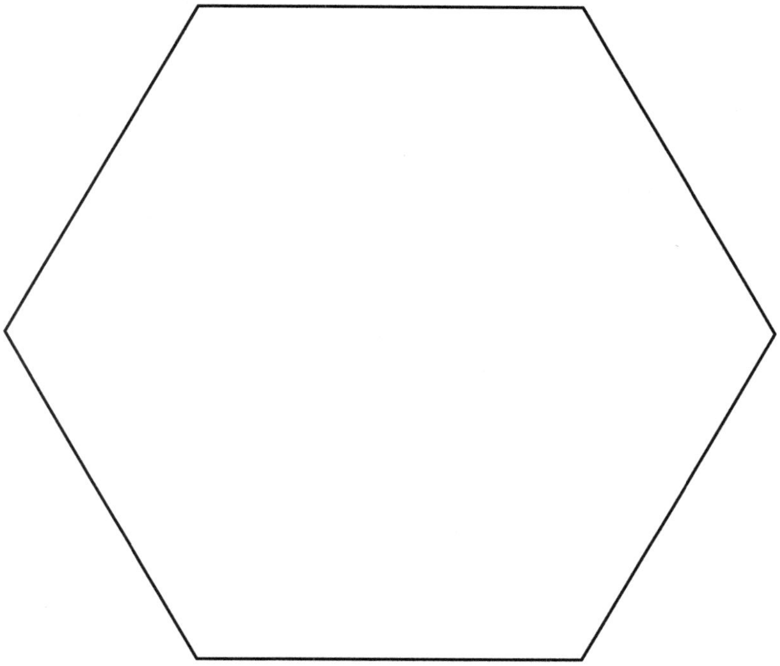

Block: ______________
Predict: ______________
Count: ______________

Block: ______________
Predict: ______________
Count: ______________

Block: ______________
Predict: ______________
Count: ______________

Build the Geoblock

1. Put Geoblocks together to build this block.

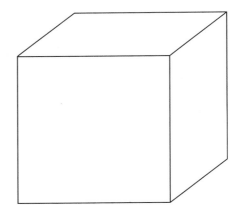

2. Put Geoblocks together to build this block.

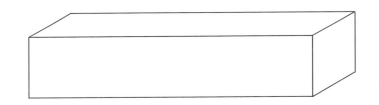

3. Put Geoblocks together to build this block.

Shapes Within Shapes

Find shapes at home that are made up of other shapes. For example, you might have a window that is a rectangle made of 4 smaller rectangles. Write about or draw pictures of at least 3 of your shapes within shapes.

Solve Puzzles Recording Sheet

Open *Shapes* and Solve Puzzles on the computer.

1. Shapes 1–4

Shape Number	Fewest Number of Blocks to Fill	What are they and how many of each block?

2. Shapes 5–10

Shape Number	Fewest Number of Blocks to Fill	What are they and how many of each block?	Number of Triangles	
			Predict	Count

Squares

Cut out the squares. Cut each square in half along
the dotted line.

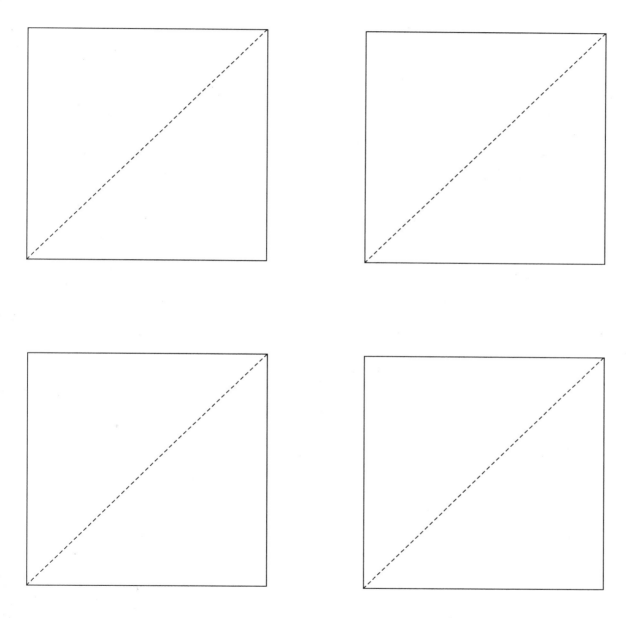

Composing New Shapes with 2 Triangles

Use 2 of the triangles from Student Sheet 10 to make new shapes. Make sure the triangles do not overlap and that they touch along at least part of one side. Trace at least 4 of your new shapes. Use the back of the paper if necessary.

Build a Building

Use interlocking cubes to build these buildings.

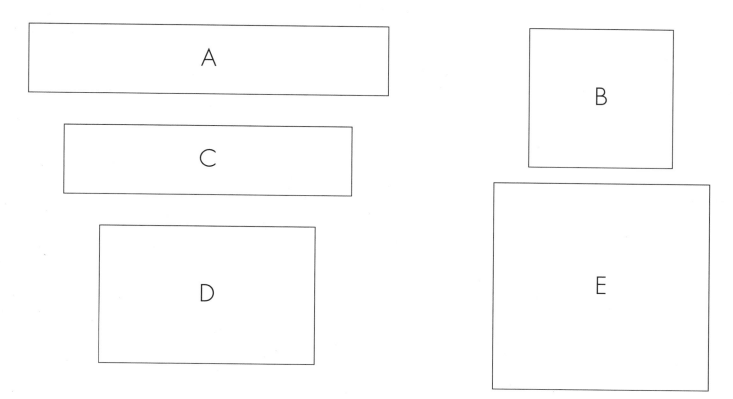

Building	Number of Rooms					
	1 story	2 stories	3 stories	4 stories	5 stories	10 stories
A						
B						
C						
D						
E						
My Building						

Investigation 1 • Sessions 6–8
Shapes, Halves, and Symmetry

Composing New Shapes with 3 or 4 Triangles

Use 3 or 4 of the triangles cut from Student Sheet 10 to make new shapes. Remember, the triangles may not overlap and they touch along at least part of one side. Trace at least 3 of your new shapes. Use the back of the paper if necessary.

Looking for Quadrilaterals (4-Sided Figures)

Look around your home or neighborhood for rectangles and other quadrilaterals. You can also look in magazines for pictures of things that are 4-sided shapes.

If you can, bring some of your shapes to school. If you find quadrilaterals that you can't take to school, draw them here. Use the back of the paper if necessary.

Which One Has the Most?

Look at the rectangles on Which Is Biggest? sheets.

Pretend that rectangles A, B, C, D, E, F, and G are chocolate bars.

1. Which chocolate bar would you want?_____
   Why?

   _____

2. Which chocolate bar has the most chocolate?_____
   How do you know?

   _____

3. Which chocolate bar has the least chocolate?_____
   How do you know?

   _____

4. Are there any chocolate bars that have the same amount of
   chocolate?_____
   If so, which ones?_____
   How do you know?

Inch Graph Paper

Investigation 2 • Session 3
Shapes, Halves, and Symmetry

Only One Rectangle

Cut carefully on the dotted lines. Use the squares you cut out to make rectangles. Which numbers between 1 and 12 make only one rectangle?

What did you notice or discover while you made your rectangles?

Growing Rectangles

Open *Shapes* and Create a Tiling.

1. Make a row or column of squares.

2. Count the number of squares in one row or column.

3. Click on the Glue tool . Click on each square.

4. Click on the Duplicate tool . Click on one row or column.

5. Place below or beside by dragging.

6. Click on the Pattern tool to make the tiling row.

| | Total Squares in Rectangle | |
	Predict	Count
1 row or column		
2 rows or columns		
3 rows or columns		
4 rows or columns		

Rectangle Riddles (1–3)

Use color tiles to make a solution to each riddle.
Draw a picture of your solution.

1. This rectangle has 5 rows.
It has 25 tiles.

2. This rectangle has 3 rows.
There are 7 tiles in each row.

3. This rectangle has 12 tiles.

Rectangle Riddles (4-6)

Use color tiles to make a solution to each riddle.
Draw a picture of your solution.

4. This rectangle has 3 columns and 3 rows.

_____

5. This rectangle has 15 tiles.
     It has 3 columns.

_____

6. Write a rectangle riddle of your own.

Cut out the rectangles. Which is the smallest? Which is the biggest?
Put them in order.

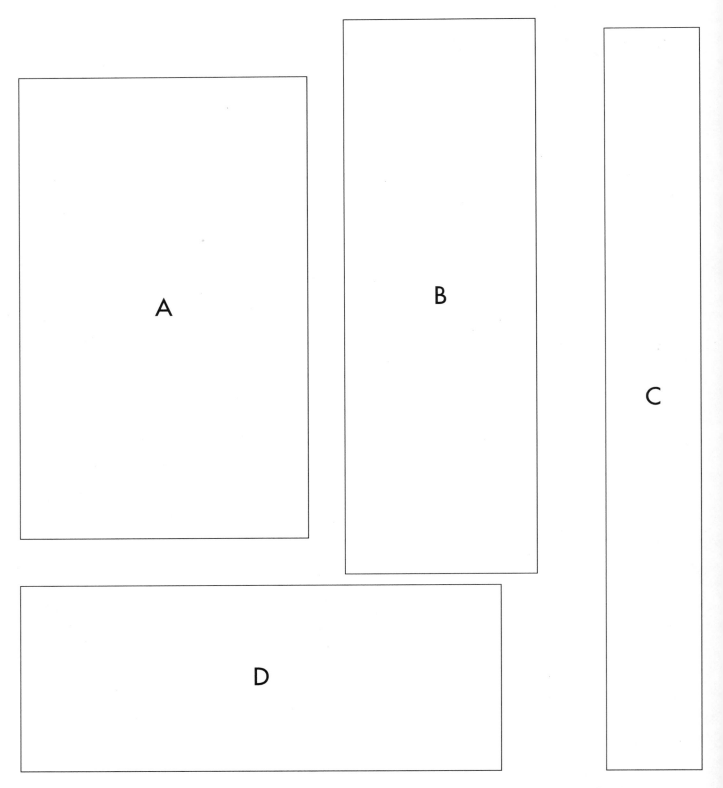

Cut out the rectangles. Which is the smallest? Which is the biggest? Put them in order.

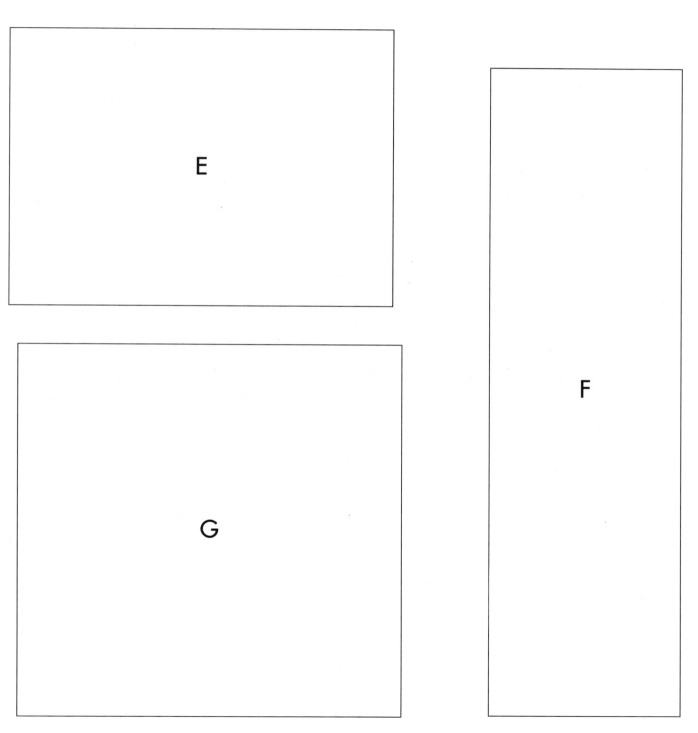

191

Halves and Not Halves

Use 2 colors. Color each rectangle to show halves and not halves.

1.   Half-and-Half   Not Half-and-Half

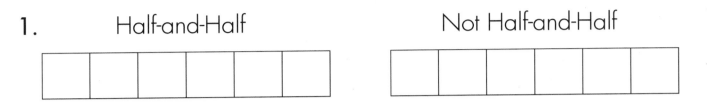

2.   Half-and-Half   Not Half-and-Half

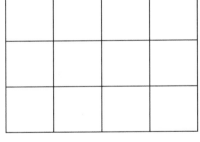   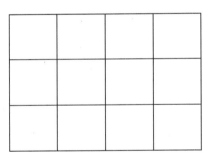

3.   Half-and-Half   Not Half-and-Half

4.   Half-and-Half   Not Half-and-Half

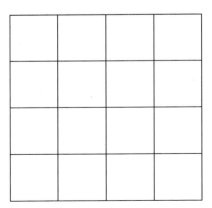

Things That Come in Halves

Find things around your home that come in halves.
Draw a picture of any items you find. Bring one or
two things to school if you can.

Designing Shapes That Can Be Cut in Half

Design shapes that can be cut into two equal halves. Can you design a shape that can be cut into two equal halves more than one way? You might want to cut these squares out and fold them or cut your designs in half to check.

Half-and-Half Flags

Make a rectangular flag on Student Sheet 16, Inch Graph Paper. Half of the squares in your flag need to be one color, and the other half needs to be a different color.

I used _____ squares.

One-half is _____ color.

One-half is _____ color.

Each half has _____ squares.

196

Looking for Symmetry

Look in and around your home for things that are symmetrical. You might find symmetrical pictures or designs in magazines or catalogs. Bring any items you find to school if you can, or draw a picture of them here.

Exploring Mirror Symmetry

Use a mirror at home to find objects that have mirror symmetry. Try many different kinds of things. Describe the most interesting object you find using words or pictures. Use the back of this sheet if necessary.

_____

_____

_____

_____

_____

Fold and Cut

Fold one piece of paper in half. Cut a shape on the fold. Do this as many times as you like.

Bring one shape and the paper you cut it out of to school.

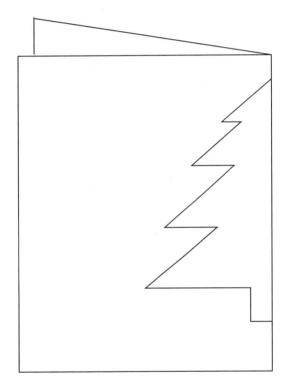

Extension

Can you figure out a way to create shapes with two or more lines of symmetry?

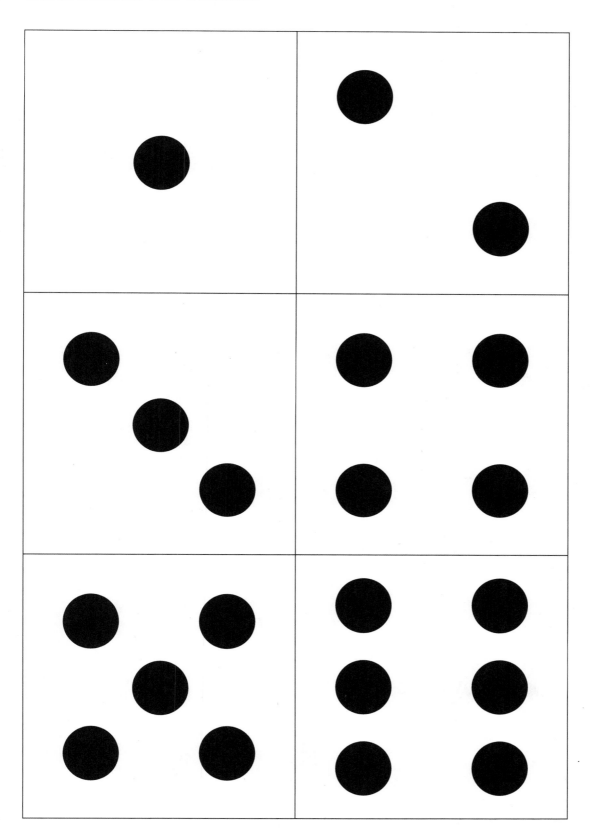

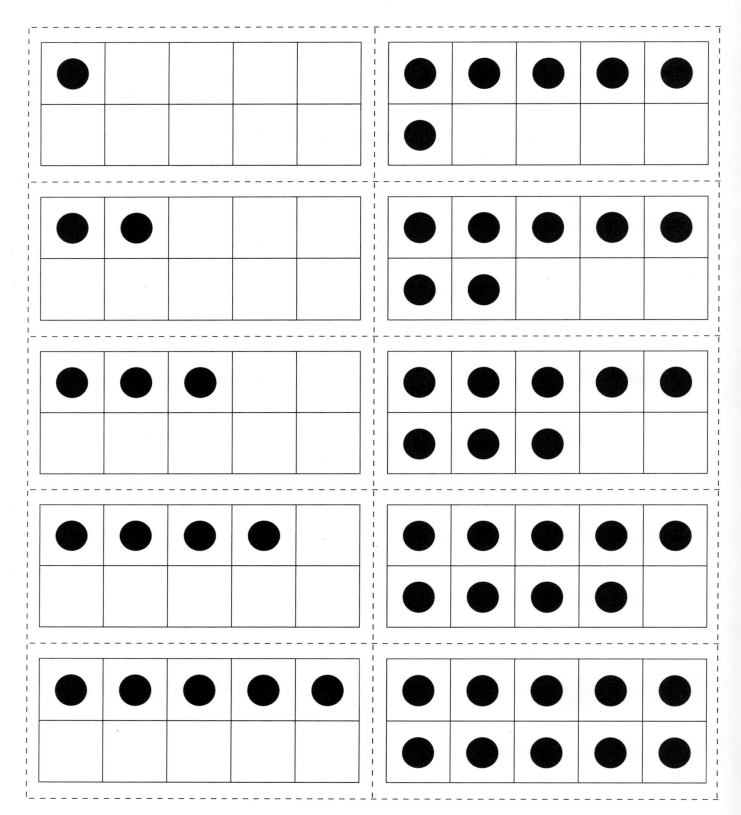

Unit Resource
Shapes, Halves, and Symmetry

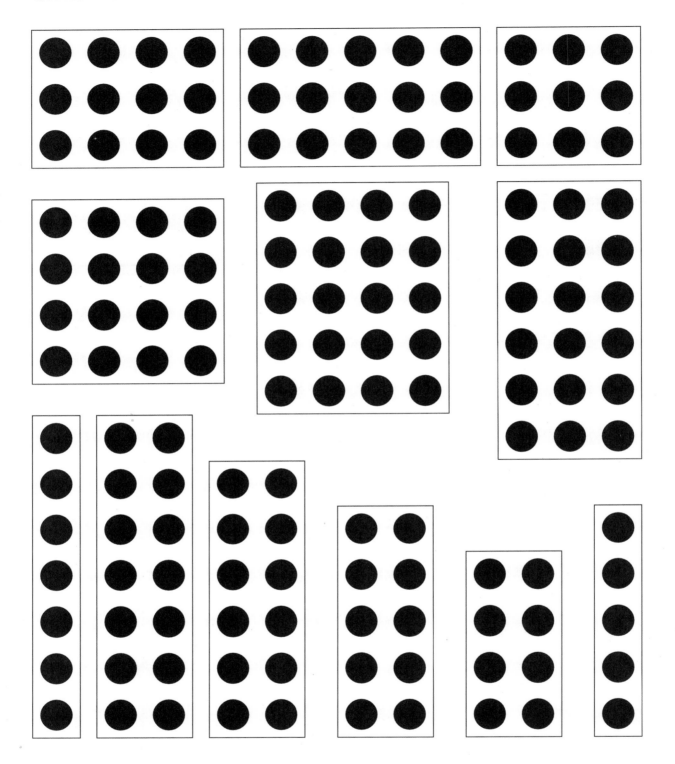

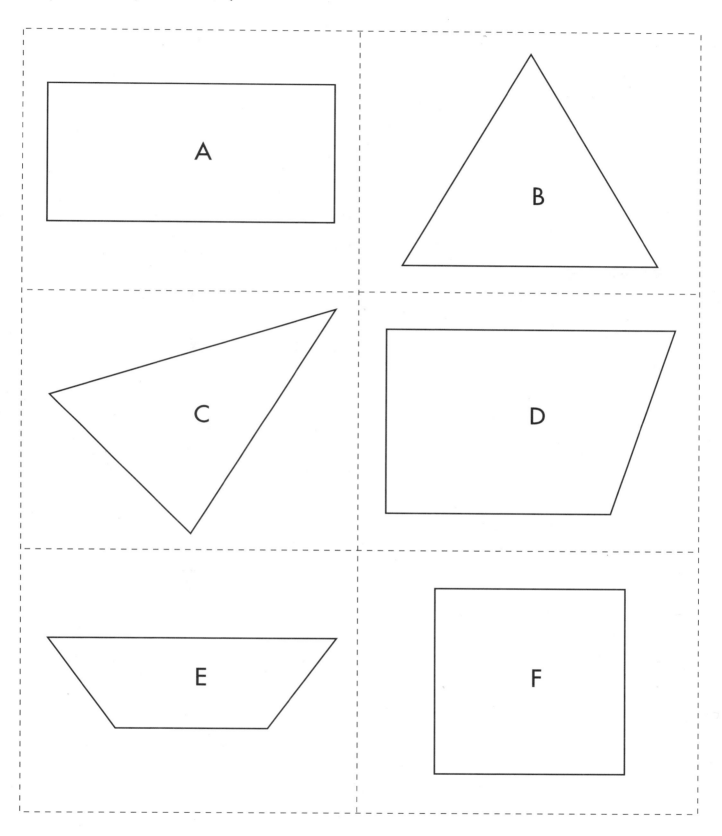

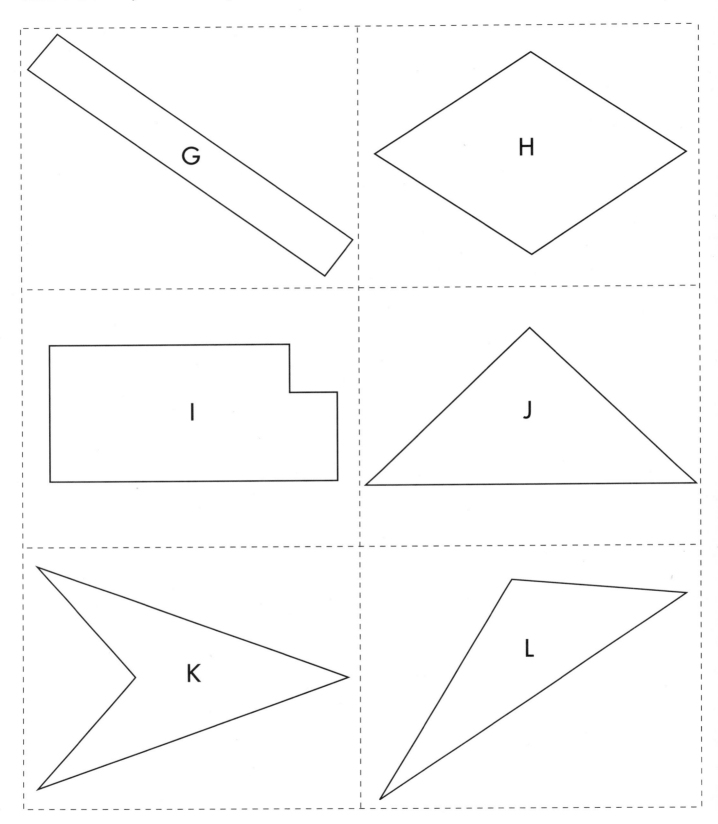

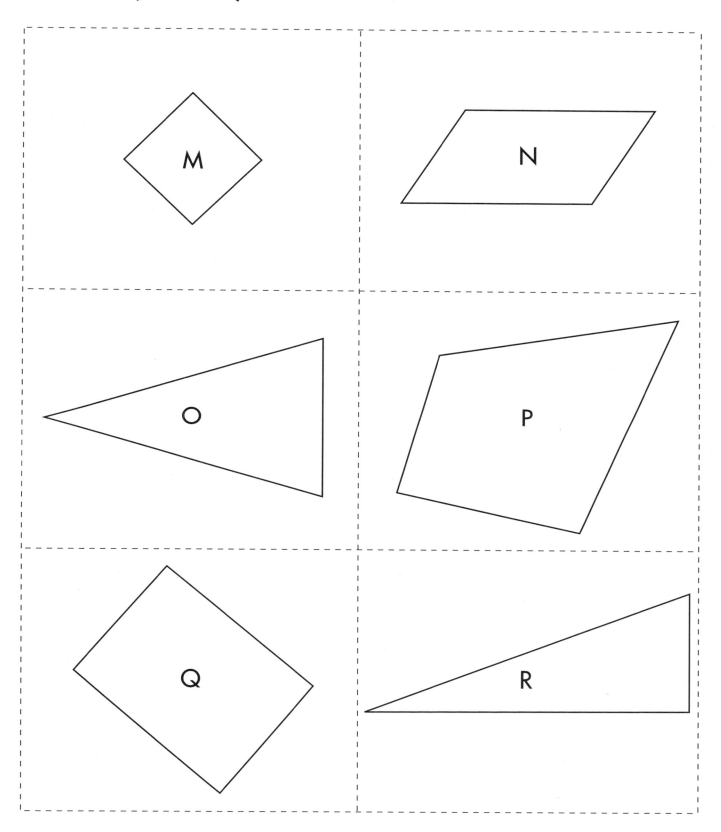

Practice Pages

This optional section provides homework ideas for teachers who want or need to give more homework than is assigned to accompany the activities in this unit. The problems included here provide additional practice in learning about number relationships and in solving computation and number problems. For number units, you may want to use some of these if your students need more work in these areas or if you want to assign daily homework. For other units, you can use these problems so that students can continue to work on developing number and computation sense while they are focusing on other mathematical content in class. We recommend that you introduce activities in class before assigning related problems for homework.

Turn Over 10   Students play this game in the units *Mathematical Thinking at Grade 2* and *Coins, Coupons, and Combinations*. If your students are familiar with the game, you can simply send home the directions and Number Cards so that students can play at home. If your students have not played the game before, introduce it in class and have students play once or twice before sending it home. You might have students do this activity four times for homework in this unit.

Close to 20   Students are introduced to this game in *Coins, Coupons, and Combinations*. If your students are familiar with the game, you can simply send home the directions, score sheet, and Number Cards so that students can play at home. If your students have not played the game before, introduce it in class and have students play once or twice before sending it home. You might have students do this activity four times for homework in this unit.

Number Strings   This type of problem is introduced in the unit *Coins, Coupons, and Combinations*. Provided here are three sheets of problems. You can also make up other problems in this format, using numbers that are appropriate for your students. For each sheet, students solve the problems and then record their strategies, using words, pictures, or number sentences.

Turn Over 10

Materials: Deck of Number Cards 0–10 (four of each) plus four wild cards

Players: 2 to 3

How to Play

The object of the game is to turn over and collect combinations of cards that total 10.

1. Arrange the cards face down in four rows of five cards. Place the rest of the deck face down in a pile.

2. Take turns. On a turn, turn over one card and then another. A wild card can be made into any number.

   If the total is less than 10, turn over another card.

   If the total is more than 10, your turn is over and the cards are turned face down in the same place.

   If the total is 10, take the cards and replace them with cards from the deck. You get another turn.

3. Place each of your card combinations of 10 in separate piles so they don't get mixed up.

4. The game is over when no more 10's can be made.

5. At the end of the game, make a list of the number combinations for 10 that you made.

Close to 20

Materials: Deck of Number Cards 0–10 (four of each) with the wild cards removed; Close to 20 Score Sheet; counters

Players: 2

How to Play

The object of the game is to choose three cards that total as close to 20 as possible.

1. Deal five cards to each player.

2. Take turns. Use any three of your cards to make a total that is as close to 20 as possible.

3. Write these numbers and the total on the Close to 20 Score Sheet.

4. Find your score. The score for the round is the difference between the total and 20. For example, if you choose $8 + 7 + 3$, your total is 18 and your score for the round is 2.

5. After you record your score, take that many counters.

6. Put the cards you used in a discard pile and deal three new cards to each player. If you run out of cards before the end of the game, shuffle the discard pile and use those cards again.

7. After five rounds, total your score and count your counters. These two numbers should be the same. The player with the lowest score and the fewest counters is the winner.

Close to 20 Score Sheet

PLAYER 1   SCORE

Round 1: _____ + _____ + _____ = _____   _____

Round 2: _____ + _____ + _____ = _____   _____

Round 3: _____ + _____ + _____ = _____   _____

Round 4: _____ + _____ + _____ = _____   _____

Round 5: _____ + _____ + _____ = _____   _____

                                    TOTAL SCORE   _____

PLAYER 2   SCORE

Round 1: _____ + _____ + _____ = _____   _____

Round 2: _____ + _____ + _____ = _____   _____

Round 3: _____ + _____ + _____ = _____   _____

Round 4: _____ + _____ + _____ = _____   _____

Round 5: _____ + _____ + _____ = _____   _____

                                    TOTAL SCORE   _____

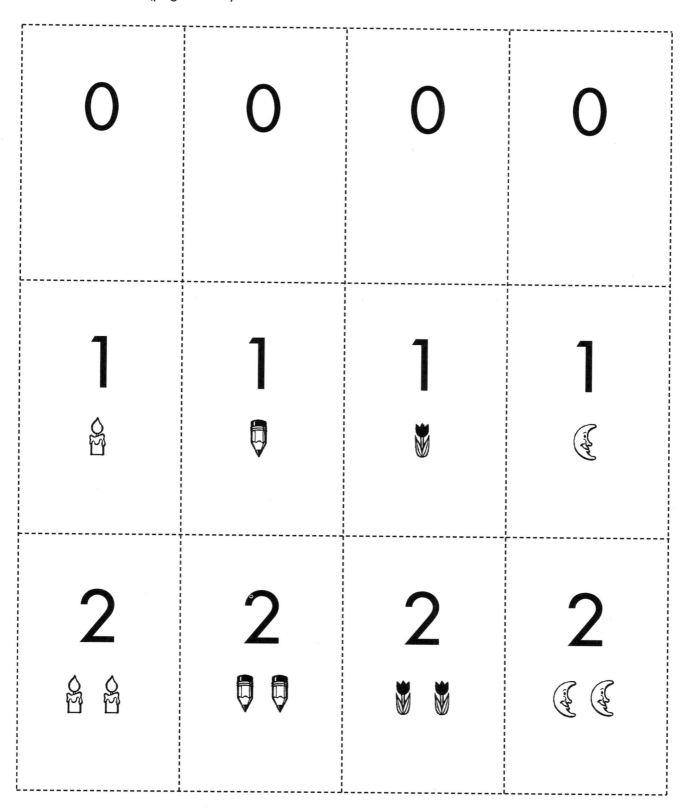

Practice Page
Shapes, Halves, and Symmetry

3	**3**	**3**	**3**
4	**4**	**4**	**4**
5	**5**	**5**	**5**

214

6	6	6	6
7	7	7	7
8	8	8	8

215

Practice Page
Shapes, Halves, and Symmetry

9	**9**	**9**	**9**
10	**10**	**10**	**10**
Wild Card	Wild Card	Wild Card	Wild Card

216

Practice Page A

Show how you solved each problem. You can use words, pictures, or number sentences.

7 + 9 + 3 + 4 =	8 + 7 + 1 + 2 =
5 + 3 + 2 + 7 =	6 + 5 + 4 + 3 =
1 + 8 + 2 + 6 + 9 =	3 + 9 + 5 + 2 =
7 + 4 + 5 + 3 =	3 + 8 + 5 + 7 + 2 =

Practice Page
Shapes, Halves, and Symmetry

Practice Page B

Show how you solved each problem. You can use words, pictures, or number sentences.

6 + 5 + 2 + 7 =	9 + 8 + 1 + 2 =
4 + 3 + 3 + 8 =	1 + 5 + 9 + 3 =
2 + 4 + 6 + 8 =	3 + 8 + 2 + 9 + 4 =
6 + 4 + 5 + 3 + 5 =	2 + 9 + 6 + 1 + 8 =

Practice Page C

Show how you solved each problem. You can use words, pictures, or number sentences.

$2 + 7 + 8 + 3 + 7 =$	$3 + 7 + 1 + 2 + 9 =$
$4 + 3 + 5 + 7 + 5 =$	$7 + 6 + 5 + 4 + 3 =$
$2 + 8 + 4 + 7 + 6 =$	$1 + 9 + 5 + 2 + 8 =$
$1 + 2 + 3 + 8 + 9 + 7 =$	$5 + 2 + 5 + 7 + 1 + 5 =$